Aurivillius Phase Materials

Exploring lead-free ferroelectrics

Series Editor

R Barry Johnson, a Senior Research Professor at Alabama A&M University, has been involved for over 50 years in lens design, optical systems design, electro-optical systems engineering, and photonics. He has been a faculty member at three academic institutions engaged in optics education and research, employed by a number of companies, and provided consulting services.

Dr Johnson is an IOP Fellow, SPIE Fellow and Life Member, OSA Fellow, and was the 1987 President of SPIE. He serves on the editorial board of *Infrared Physics & Technology* and *Advances in Optical Technologies.* Dr Johnson has been awarded many patents, has published numerous papers and several books and book chapters, and was awarded the 2012 OSA/SPIE JosephWGoodman Book Writing Award for *Lens Design Fundamentals*, (second edition). He is a perennial co-chair of the annual SPIE Current Developments in Lens Design and Optical Engineering Conference.

Foreword

Until the 1960s the field of optics was primarily concentrated in the classical areas of photography, cameras, binoculars, telescopes, spectrometers, colorimeters, radiometers, etc. In the late 1960s, optics began to blossom with the advent of new types of infrared detectors, liquid crystal displays (LCDs), light emitting diodes (LEDs), charge coupled devices (CCDs), lasers, holography, fiber optics, new optical materials, advances in optical and mechanical fabrication, new optical design programs, and many more technologies. With the development of the LED, LCD, CCD and other electro-optical devices, the term ‘photonics’ came into vogue in the 1980s to describe the science of using light in the development of new technologies and the performance of a myriad of applications. Today, optics and photonics are truly pervasive throughout society and new technologies are continuing to emerge. The objective of this series is to provide students, researchers, and those who enjoy self-teaching with a wideranging collection of books that each focus on a relevant topic in the technologies and applications of optics and photonics. These books will provide knowledge to prepare the reader to be better able to participate in these exciting areas now and in the future. The title of this series is *Emerging Technologies in Optics and Photonics* where ‘emerging’ is taken to mean ‘coming into existence’, ‘coming into maturity’, and ‘coming into prominence’. IOP Publishing and I hope that you find this series of significant value to you and your career.

Aurivillius Phase Materials

Exploring lead-free ferroelectrics

Rajnish Kurchania
Department of Physics, Maulana Azad National Institute of Technology, Bhopal, India

Oroosa Subohi
Department of Physics, Visvesvaraya National Institute of Technology, Nagpur, India

IOP Publishing, Bristol, UK

ISBN 978-0-7503-3419-8 (ebook)
ISBN 978-0-7503-3417-4 (print)
ISBN 978-0-7503-3420-4 (myPrint)
ISBN 978-0-7503-3418-1 (mobi)

DOI 10.1088/978-0-7503-3419-8

Version: 20211201

IOP ebooks

British Library Cataloguing-in-Publication Data: A catalogue record for this book is available from the British Library.

Published by IOP Publishing, wholly owned by The Institute of Physics, London

IOP Publishing, Temple Circus, Temple Way, Bristol, BS1 6HG, UK

US Office: IOP Publishing, Inc., 190 North Independence Mall West, Suite 601, Philadelphia, PA 19106, USA

Dedicated to

My parents Mr Ramesh Chandra and Mrs Kusum, wife Vandana, and daughters Riya and Shreya.

— Rajnish Kurchania

My parents (late) Mr Mohd. Kasim Raza and Mrs Soofia, husband Shams, and daughter Aleena.

— Oroosa Subohi

Contents

Preface

This book aims to provide an overview of a new class of lead-free ferroelectric compounds namely *Aurivillius phase materials*. The emphasis is laid on their properties owing to their layered structure and their potential applications in ferroelectric random access memory (FeRAM), capacitors, sensors, transducers, actuators, and other electronic devices. Most of the literature available on ferroelectrics deals with all types of ferroelectric materials in general. This book discusses explicitly the materials belonging to the Aurivillius family. Amongst the different types of ferroelectric materials, Aurivillius phase materials present distinctive properties due to their unique structure and they also possess a high Curie temperature (>300 °C). There are a wide range of materials which belong to this family and can be subdivided into many more categories according to the number of layers that they possess. This book addresses the need to address exclusively this family of layered ferroelectrics in detail.

The classification of the family of Aurivillius phase materials based on the number of perovskite layers present is discussed in this book. The relationship between the structure and properties of this family of compounds is analysed in accordance with their classification. The frequency and temperature dependent properties (electrical, dielectric, ferroelectric, and piezoelectric) of these compounds as a function of the number of layers is also presented. A cohesive picture of the electrical, ferroelectric, and dielectric properties of Aurivillius compounds based on the different synthesis routes adopted for their preparation is presented. In addition, modification of the properties of these compounds through the substitution of cations by isovalent and heterovalent ions is realized. A comprehensive review of the work done by various researchers in this field in addition to our own work is presented. Hence this book will provide insight into the possibilities of new lead-free ferroelectric materials that can be used in various device applications with the scope to further promote research in this exciting area of environmentally friendly compounds.

The basic concepts, principles, and methods for measuring the physical properties of ferroelectric materials are introduced in the first chapter for those readers for whom ferroelectricity is a new subject. Therefore this book is suitable for researchers, scientists and engineers who are working in the field of ferroelectric and piezoelectric materials and their various applications.

Rajnish Kurchania and Oroosa Subohi

Foreword

Aurivillius phases, or bismuth layer structure materials, as they are sometime known, are both scientifically fascinating and potentially technologically important. Despite their often bewilderingly complex compositions and large unit cells, their crystal structures are relatively simple when viewed as a series of rock salt-like bismuth oxide layers interleaved with perovskite stacks. This structure provides a range of useful functional properties, including ferromagnetism, ferroelectricity, piezoelectricity, magnetoelectricity, and photo-catalysis. As with the simple perovskites, variation in composition allows for tailoring of properties. For example, the Aurivillius ferroelectric Curie temperature is dependent on the relative ion sizes in the perovskite stacks and, in general, can be much higher than the equivalent simple perovskites. Plus, Aurivillius phases offer the additional variable of the ratio of the number of perovskite cells to bismuth layers. This can be manipulated in a highly systematic way, providing an extra lever with which to engineer their properties. Moreover, the large crystalline anisotropy of layer structures can often be exploited in both thin film and ceramic forms to produce materials with highly preferred crystallographic orientation or texture and thereby further enhance the anisotropic properties.

This book acts as an invaluable primer and guide to this class of materials, and is suitable for both beginners and relative experts. Its focus is on the ferroelectric and related properties of the materials and in the opening chapter provides a concise introduction to, or reminder of, the phenomena and archetypal materials associated with ferroelectricity, before introducing the layer structure oxides themselves. The subsequent three chapters address the structure and properties of typical examples of Aurivillius phases, categorizing them according to the parity of the number of perovskite layers per bismuth oxide block. These chapters may be considered a starting point for new researchers, providing sufficient understanding of how to formulate and characterise materials that can be extended to address novel examples of the class. The final chapter focuses on current and potential applications of the materials. Each chapter is supported by numerous references to the key literature in the field.

The book is timely (although may also be considered to be long overdue!), as the search for lead-free alternatives to the market-leading lead-based piezoelectric and ferroelectric materials has all but exhausted the options in the simple perovskites. Aurivillius phases offer valuable alternatives, particularly where a high Curie temperature is required, or the relative lack of domain-related aging is advantageous. In some cases, notably in the search for improved magnetoelectric materials, the ability to mix odd and even numbers of layers may provide opportunities to engineer properties relying on combinations of cations that may be incompatible in the simple perovskites.

Both authors are experts in the field of layer structure materials, having undertaken research on a range of Aurivillius phases for a variety of applications. They are experienced in both the fabrication and characterisation of the materials as ceramics and thin films and bring to the subject the insights of academic researchers, coupled with an understanding of the requirements for commercial exploitation.

For new researchers in the field, or for those just wishing to broaden their knowledge to include layer structure ferroelectrics, this book provides an excellent starting point and guide with which to begin your journey. I make this statement based on personal experience, as I recently had need to take the plunge and propose a programme of research in the field of Aurivillius phases. With uncanny timing, the manuscript of this book arrived in my inbox for review and in a short time has proved to be a vital source in constructing that research proposal.

Thank you to Professor Kurchania and Dr Subohi for making my life a little easier and I promise not to hold you responsible should my research proposal fail to be funded!

Professor Andrew J Bell FREng FIMMM FIEEE
School of Chemical and Process Engineering, University of Leeds, UK
August 2021

Acknowledgement

We are grateful to our host institutions (MANIT, Bhopal, and VNIT, Nagpur) and our colleagues for their constant support. We are obliged to Professor Andrew J Bell, University of Leeds, UK for writing an excellent introduction for this book through his foreword. We are indebted to Professor Steven J Milne, University of Leeds, UK, Professor Chris R Bowen University of Bath, UK, Dr Richard J Ball, University of Bath, UK, Professor D C Dube, IIT Delhi, Professor S C Mathur, IIT Delhi, Professor M M Malik, MANIT Bhopal and Professor G S Kumar, Osmania University, Hyderabad, India, for various fruitful discussions. We are thankful to Ms Sravani Sarkar for proofreading the manuscript of this book. We thank Ms Ashley Gasque Stec, Senior Commissioning Editor, IOP Publishing, and Mr Robert Trevelyan, Editorial Assistant, IOP Publishing, for their help. We appreciate Affan for typesetting certain parts of manuscript.

Author biographies

Rajnish Kurchania

Dr Rajnish Kurchania is a Fellow of the Institute of Materials, Minerals and Mining (FIMMM) UK, Chartered Engineer (CEng) and a Chartered Scientist (CSci). He is currently working as a Professor in the Department of Physics at Maulana Azad National Institute of Technology (MANIT), Bhopal, India. He has held various administrative responsibilities such as Associate Dean (PG and PhD) and Head of the Department. His earlier appointments include Reader, Associate Professor in the Physics Department at RGPV, Bhopal, Senior Project Leader (Technology) with The Welding Institute (TWI), Port Talbot, UK, Post Doctoral Research Fellow at the University of Leeds, UK, and the University of Bath, UK. He established a University Teaching Department (UTD) in the area of Nanotechnology at the Rajiv Gandhi Proudyogiki Vishwavidyalaya (RGPV) Bhopal, and served as the Founder Head of the School of Nanotechnology RGPV. He received his PhD in Materials Science and Engineering from the University of Leeds in 1998 and was awarded his MSc in Physics from Barkatullah University, Bhopal in 1991. He is a recipient of the Overseas Research Student (ORS) Award from the Committee of Vice-Chancellors and Principals (CVCP) of the Universities of the UK and the Tetley-Lupton Scholarship from the University of Leeds to support his doctoral studies at the University of Leeds. He was appointed as Senior Visiting Research Fellow by the Faculty of Engineering and Design at the University of Bath, UK. He has authored over 100 peer reviewed research papers in international journals and conferences of repute. He has guided twelve PhD and fifty-six MTech students so far. His research interests include advanced functional materials, ferroelectric thin films, photocatalytic materials, solar photovoltaics, non-destructive testing, and computational materials science. He is a life member of the Materials Research Society of India (MRSI), the Indian Society for Technical Education (ISTE), and Vigyan Bharti (VIBHA). He has established valuable working relationships with various academic institutions and industries in UK, USA, India and Russia. He received funding and successfully completed a number of externally funded research projects from prestigious funding agencies such as: Going Global Partnerships Collaborative Grant funded by the British Council, project partners: University of Bath UK, MANIT Bhopal, IIT Madras, SPAs, IIT Roorkee, India; DST-SERB Delhi, India; M. P. Council of Science and Technology Bhopal, India; the Royal Academy of Engineering (RAE), London, UK; the British Council UK–India Education Research Initiative (UKIERI phase-II) for collaborative research with the University of Bath and IIT Delhi; the Russian Foundation of Basic Research and the Department of Science and Technology (RFBR-DST) in collaboration with JEC Jaipur and the Russian Academy of Science, Moscow.

Oroosa Subohi

Dr Oroosa Subohi completed her post-graduate degree in Physics at Osmania University, Hyderabad, India, in 2000 and was awarded her PhD (Physics) by the Maulana Azad National Institute of Technology, Bhopal, India, in 2015. She has authored more than 25 articles on ferroelectric and multiferroic ceramics in journals of international repute and has contributed an invited book chapter. She has guided five students for their master's dissertation and is currently supervising PhD scholars. She is a life member of the Materials Research Society of India. Currently, she is working as an Assistant Professor in the Department of Physics, Visvesvaraya National Institute of Technology, Nagpur, India.

IOP Publishing

Aurivillius Phase Materials
Exploring lead-free ferroelectrics
Rajnish Kurchania and Oroosa Subohi

Chapter 1

Introduction

1.1 General introduction

Materials scientists have long been interested in materials engineering, i.e. modifying existing materials and the synthesis of new materials to arrive at desired properties which can be tuned to a specific activity. Currently, materials for different applications are available in different forms such as single crystals, polycrystallines, and thin or thick films. Of these materials, ceramics are of current interest since they can be used as electrical conductors (e.g. for humidity sensors or fuel cells), semiconductors for energy conversion, and insulators for dielectric, ferroelectric, pyroelectric, or piezoelectric applications [1, 2]. The materials that have been specially designed for specific electrical, magnetic, and optical applications are now classified under electroceramics as distinct from other types such as advanced structural ceramics.

The sub-classes of electroceramics have grown in a parallel manner, for example, ferroelectrics for high dielectric capacitors and non-volatile memories, ferrites for data and information storage, solid electrolytes for energy storage and conversion, piezoelectrics for sonar, semiconducting oxides for environmental monitoring, and a host of others. Micro-sensors and micro-actuators based on electroceramics are being fabricated in which ceramic materials and micro-technology are combined to form efficient micro-devices for production control, environment monitoring, and biomedical applications.

The history of ferroelectricity began about 80 years ago, with the discovery of Rochelle salt (sodium potassium tartarate tetrahydrate ($NaKC_4H_40_6.4H_20$)), which shows the peculiar phenomenon of 'ferroelectricity'. Valesek [3] observed this phenomenon in Rochelle salt for the first time from the measurement of its dielectric properties. Initially it was thought that this peculiar phenomenon in Rochelle salt is due to its complicated structure, but this impression vanished gradually with the discovery of similar phenomenon in a series of other compounds. The first

doi:10.1088/978-0-7503-3419-8ch1

systematic attempt to compile a list of ferroelectrics was carried out by Jona and Shirane [4].

The name 'ferroelectricity' is a consequence of the phenomenological similarity with ferromagnetism. There is no relation between ferromagnetism and ferroelectricity. The development of ferroelectric materials in order to improve the ferroelectric properties led to widespread industrial and commercial applications. In the race of device miniaturization, smart ceramic materials are technologically significant, particularly in sensors, actuators, transducers, memory devices, and capacitors.

In the early 1950s, most of the applications in the field of capacitors and piezoelectric sensors used only $BaTiO_3$ (BT) based ceramics. Thereafter, various other ferroelectric ceramics including lead titanate ($PbTiO_3$), lead zirconate titanate (PZT), lead lanthanum zirconate titanate (PLZT), and relaxor ferroelectrics such as lead magnesium niobate (PMN) were developed and utilized in various device applications that can be credited directly to this most unusual phenomenon [5]. Due to the excellent piezoelectric properties of lead-based ceramics, they are the most commonly used piezoelectrics and are used widely in ferroelectric random access memory (FRAM), actuators, and sensors.

However, as lead-based ceramics contain lead—a highly toxic substance—as a part of their chemical composition, it leads to environmental hazards. Thus, there is a requirement for lead-free ferroelectric materials with high performance that can substitute lead-based ferroelectric ceramics.

Therefore, alternative lead-free materials, namely bismuth layer structured ferroelectrics (BLSF) belonging to the Aurivillius family, are being prepared for advanced memory applications. Moreover, these materials are compatible with CMOS technology [6]. BLSF materials are also referred to as future promising smart memory materials. These bismuth based ferroelectric materials present the following important properties:

(i) High Curie temperature (300 °C–900 °C).
(ii) High anisotropy.
(iii) Fatigue free nature.
(iv) Lower crystallization temperature.
(v) Lead-free nature.
(vi) Good piezoelectric and pyroelectric response.

The properties of ferroelectrics depend largely on the domain structure, the nature of interaction between the domain boundaries, and various sorts of defects. Detailed impedance and conductivity analysis is very sensitive to the motion of defects and domain walls and is useful in understanding the effect of oxygen vacancies and the pinning of domain walls. Therefore, these investigations are very important in improving the ferroelectric properties of materials.

1.2 Materials classification on the basis of symmetry

The symmetry of different materials, such as single crystals, thin films, and polycrystalline or amorphous materials, influences their physical properties.

According to Neumann's principle, the external properties of a material are dependent on the internal crystal structure symmetry [7]. Four symmetry operations are employed to understand the symmetry about a point in space. They are: (i) the centre of symmetry, (ii) the axes of symmetry, (iii) the plane of symmetry, and (iv) a combination of the above [8]. Based on these symmetry operations solids are divided into a total of 32 crystal classes (point groups).

Dielectric materials belong to any one of the 32 crystal classes (point groups). Of these, 11 are centrosymmetric, i.e. they possess a centre of symmetry, and 21 are non-centrosymmetric. Among the non-centrosymmetric point groups, twenty groups are piezoelectric, in which the polarization can be induced with the application of mechanical stress. Out of these 20 piezoelectric point groups, ten point groups belong to the category of polar materials, which possess spontaneous polarization even in the absence of any applied electric field or stress. They are called pyroelectrics, as the spontaneous polarization in these materials is detected by heating these samples. The asymmetry inside the elementary crystal unit cell is the key reason for spontaneous polarization in the case of polar materials. Those polar materials whose direction of spontaneous polarization can be altered by the application of an external electric field are known as *ferroelectrics*. The classification of materials based on symmetry operations is shown in figure 1.1 and the flow chart for the same is shown in figure 1.2.

1.3 The basics of ferroelectricity and related phenomenon

1.3.1 Dielectric materials

Dielectrics are a special kind of insulating materials. As they are insulators, they possess virtually no free charge carriers. When dielectric materials are placed in an external electric field they undergo polarization. Polarization is the process of displacement of oppositely charged particles in opposite directions due to an externally applied field, which results in the formation of a permanent dipole moment. The strength of the dielectric properties of materials is measured in terms of the dielectric constant (ε) or relative permittivity (ε_r). The dielectric constant (ε) or relative permittivity (ε_r) is the measure of the ability of the material to be polarized in response to an applied electric field. Hence the greater the dielectric constant of a

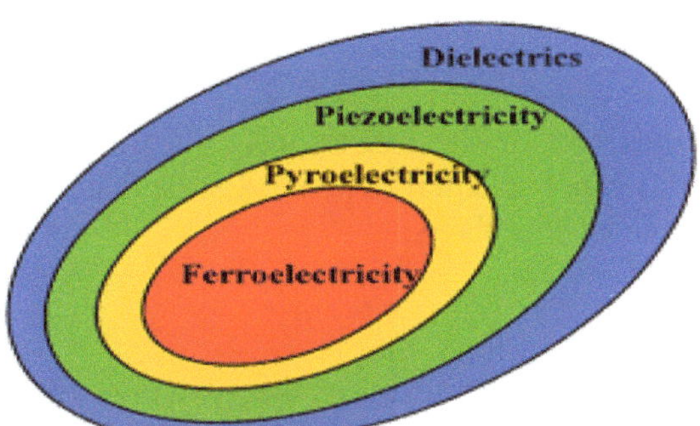

Figure 1.1. Classification of dielectric materials.

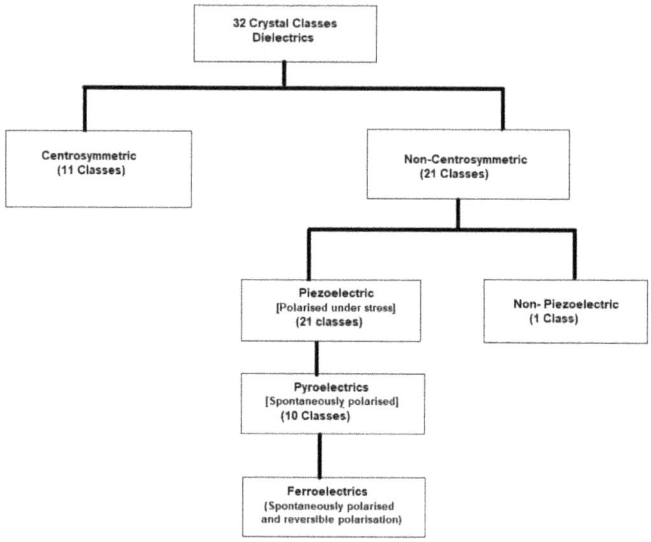

Figure 1.2. Materials classification on the basis of symmetry.

material, the greater is the polarization developed in a material for a particular applied field strength.

The energy stored in a capacitor is

$$E = \frac{1}{2}CV^2. \tag{1.1}$$

The total capacitance (C) of a capacitor is given by

$$C = \frac{\varepsilon_0 \varepsilon_r A}{d} \tag{1.2}$$

$$C_o = \varepsilon_o \frac{A}{d} \tag{1.3}$$

$$C = \varepsilon_r C_o, \tag{1.4}$$

where, ε_0 is the permittivity of the free space, ε is the permittivity or dielectric constant of the material which is equal to $\varepsilon_o{}^*\varepsilon_r$, d is the distance between the parallel plates, and A is the area of the plates:

$$\varepsilon_r{}^* = \varepsilon_r{}' - j\varepsilon_r{}'' \tag{1.5}$$

$\varepsilon_r{}^*$ represents the complex permittivity, the real part of permittivity ($\varepsilon_r{}'$) is a measure of the energy stored in a material. The imaginary part of the permittivity ($\varepsilon_r{}''$) is a measure of the energy lost in the material when an external electric field is

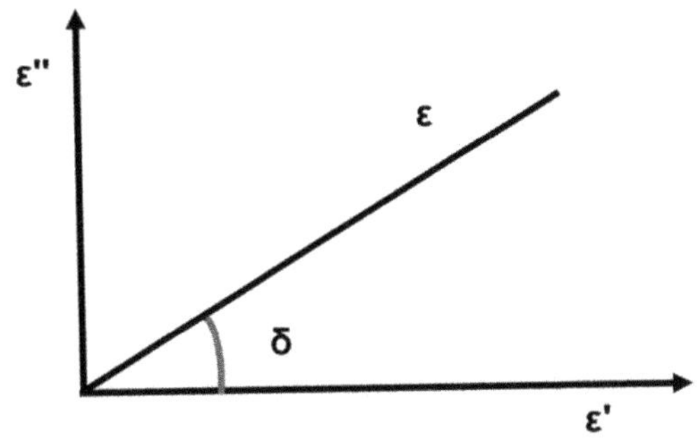

Figure 1.3. Loss tangent vector diagram.

applied. It is always greater than zero and is usually much smaller than the real part of permittivity (ε_r'). When the complex permittivity is plotted as a simple vector diagram (figure 1.3), the real and imaginary components are 90° out-of-phase. The vector sum forms an angle δ with the real axis (ε_r'). The relative 'lossiness' of a material is the ratio of the energy lost to the energy stored. The loss tangent or tanδ is defined as the ratio of the imaginary part of the dielectric constant to its real part. D denotes the dissipation factor and Q is the quality factor. The loss tangent tanδ is called the tangent loss or dissipation factor. Sometimes the term 'quality factor or Q-factor' is used with respect to an electronic microwave material, which is the reciprocal of the loss tangent. For very low-loss materials, since tan$\delta \approx \delta$, the loss tangent can be expressed in angle units, milliradians or microradians:

$$\tan\delta = \frac{\varepsilon_r''}{\varepsilon_r'}, \tag{1.6}$$

$$D = \frac{1}{Q} = \frac{\text{Energy lost per cycle}}{\text{Energy stored per cycle}}. \tag{1.7}$$

The dielectric susceptibility χ of the material is the parameter that relates the applied external electric field (E) and electrical polarization (P) by the relation

$$P = \chi\varepsilon_0 E. \tag{1.8}$$

Also, ε_r of an isotropic medium is defined by the relation

$$\varepsilon_r = \frac{(\varepsilon_0 E + P)}{\varepsilon_0 E} = 1 + \chi, \tag{1.9}$$

where $\varepsilon_0 E + P = D$ is the electric displacement field.

Therefore, the energy stored in a dielectric material is directly proportional to the dielectric constant of the material. The total polarization in a dielectric material is a combinational contribution of four different polarization mechanisms which depend on the frequency of the externally applied electric field.

The four mechanisms of polarization are the following.

When an atom is placed in an external electric field, the centre of mass of a spherical electron cloud of uniform density is displaced slightly with respect to the centre of mass of the nucleus with the charge +Ze due to the electrostatic force. This causes an induced dipole moment which results in *electronic polarization* (P_e).

When the sharing of electrons among the atoms of a molecule is not symmetrical then in such cases ions acquire partial opposite charge. When such a molecule is placed in an external electric field, the net charge tends to change the equilibrium position of the ions themselves. Thus when an external electric field is applied, the positive and negative sub-lattices are displaced in opposite directions. This gives rise to an induced dipole moment resulting in *ionic polarization* (P_i).

In some molecules, there is a permanent dipole moment even in the absence of an external electric field. This permanent dipole moment experiences a torque in the presence of an external applied electric field. This mechanism gives rise to *orientational polarization* (P_o). For orientational polarization the molecule must possess a permanent dipole moment.

In addition to these three polarization mechanisms, one more polarization mechanism, known as the *space charge polarization* mechanism, also exists. The polarization which occurs due to the accumulation of charges at the electrodes or at the interfaces is known as space charge polarization (P_s) or interfacial polarization. The main cause of space charge polarization is diffusion, in which the charges migrate over a limited distance under the influence of an applied field and forms positive and negative space charges at the grain boundaries or at the electrode–material interface. Except for P_o, all other kinds of polarization, that is P_e, P_i, and P_s, are independent of temperature. The total polarization (P_T) of a material is given by the contribution due to all types of polarizations:

$$P_T = P_e + P_i + P_o + P_s. \tag{1.10}$$

In the lower frequency region all types of polarizations (electronic, ionic, orientational, and interfacial) contribute to total polarization which gives rise to a high value of the dielectric constant. However, as the frequency increases, it becomes difficult for the dipoles to orient themselves in the direction of the applied electric field during each alternation of the electric field. Hence, a low value of dielectric constant at high frequency is obtained. The dependence of the polarization on frequency also indicates the time for which a particular mechanics of polarization remains. Electronic polarization (P_e) is the fastest and occurs completely at the instant the voltage is applied. It persists around 10^{13}–10^{15} Hz and occurs during every cycle of applied voltage. Ionic polarization (P_i) is slower compared to P_e as ions are heavier than atoms, so the time taken for displacement is greater and it occurs at frequencies between 10^9–10^{13} Hz. Orientational polarization occurs around 10^9 Hz and is slower than P_i. The slowest is the space charge polarization occurring between 50–60 Hz since it involves the diffusion of ions over several interatomic distances. Figure 1.4 represents the dependence of each polarization mechanism as a function of frequency.

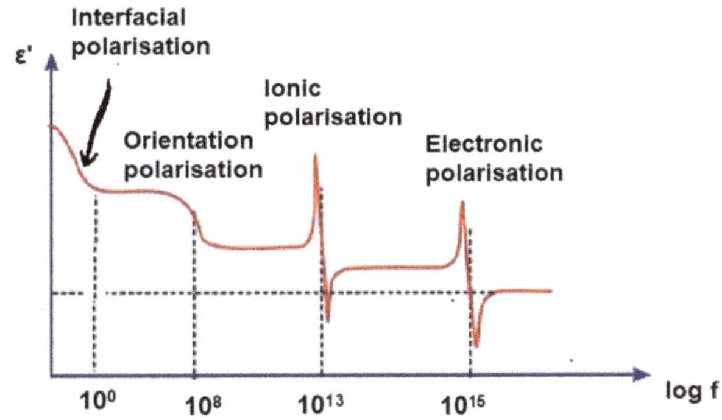

Figure 1.4. Variation of polarization with frequency.

1.3.2 Impedance spectroscopy

Complex impedance spectroscopy is a powerful tool used for studying the electrical properties of polycrystalline materials. Most of the electrical properties are affected strongly by the microstructure, and impedance spectra usually contain features that are related directly to the microstructure (the grain and grain boundaries of differing phase compositions, the suspension of one phase within another, the porosity, electrode interface effects, etc). It also enables us to evaluate the relaxation frequency $\omega_{\max}$ of the material. The relaxation frequency of the material at a given temperature is an intrinsic property of the material independent of the geometrical factors of the sample. The contribution of various processes such as grain, grain boundary, and electrode–sample interface effects, etc, can be resolved in the frequency domain. Impedance is the sum of resistance and reactance in which the latter is a function of frequency. Any material can be considered as a combination of resistance and reactance either in series or in a parallel combination of both. In the parallel mode admittance (Y) is dominant and in the series mode impedance (Z) plays a vital role. The parallel mode is more appropriate for inductive samples and the series mode for capacitive samples. Electrical insulators can be characterized using ac impedance data. There are various formalisms by which one can exploit the wealth of data. The effects which do not show up in one of the formalisms show up in another. All the four formalisms are interrelated. The different formalisms are electric modulus (M), admittance (Y), permittivity (ε), and impedance (Z), after Sinclair *et al* [10]. They are

$$M^* = \frac{1}{\varepsilon^*} = M' + iM'' \tag{1.11}$$

$$Z^* = \frac{1}{j\omega C_o \varepsilon^*} \tag{1.12}$$

$$Y^* = \frac{1}{Z^*} = \frac{j\omega C_o}{M^*}. \tag{1.13}$$

Different graphical analyses can be performed using these three equations. The general forms of data presentation are complex impedance plots (Z'' versus Z'), loss spectra (ε'' or M'' versus frequency), and the plots of ε', σ', or σ'' versus frequency. Usually, a peak is observed when a graph with these dielectric parameters is plotted with frequency. These peaks within a particular frequency range depend on the strength of relaxation. The data over a wide range of frequencies give additional information compared to fixed frequency dielectric data. Particularly, polycrystalline materials exhibit a wide variety of properties depending on the frequency. Such studies would help in understanding the dielectric nature, conduction processes, and possible relaxation phenomena. Any intrinsic property that influences the conductivity of an electrode–material system or an external stimulus can be studied by impedance measurement. The parameters derived from such a study fall into two categories (i) those relevant only to the material itself, such as conductivity, dielectric constant, etc, and (ii) those pertinent to an electrode–material interface such as the adsorption–reaction rate, capacitance of interface regions, diffusion coefficients, etc [9]. All ferroelectric materials have coupled electrical and mechanical properties. The impedance data, generally compared to an equivalent circuit, are obtained, which represent physical processes taking place in the system under investigation. An equivalent circuit is made up of ideal resistors, capacitors, inductance, and possibly various distributed circuit elements. In such a circuit, the resistance represents a conductive path and a given resistor in the circuit might account for the bulk conductivity of the material. Similarly, capacitances and inductances will be generally associated with various polarization processes. The total impedance of a polycrystalline material is the sum of the grain, grain boundary, and electrode effects [10]. By analysing the results on impedance data in more than one of the formalisms, it may be possible to separate these effects effectively.

1.3.3 Piezoelectricity

Piezoelectricity was first discovered by Jacques and Pierre Curie [11]. Piezoelectricity is defined as 'electrical polarization produced by applying mechanical strain to a crystal'. The polarization produced is directly proportional to the mechanical strain applied to the crystal. A basic requirement for a material to show piezoelectricity is that it must lack a centre of symmetry of the unit cell. This non-centrosymmetry of the unit cell creates an asymmetric charge distribution, which results in the formation of an electric dipole in the crystal as a whole. The converse is also true, i.e. when voltage is applied across the surface of the crystal the strain is produced across the surface. This strain produced in a crystal is directly proportional to the applied electric field.

The direct and converse piezoelectric effects are shown in figure 1.5 and can be expressed in tensor notation as

$$D_i = d_{ijk}\sigma_{jk} \text{ (direct effect)} \tag{1.14}$$

$$S_{ij} = d_{kij}E_k \text{ (converse effect),} \tag{1.15}$$

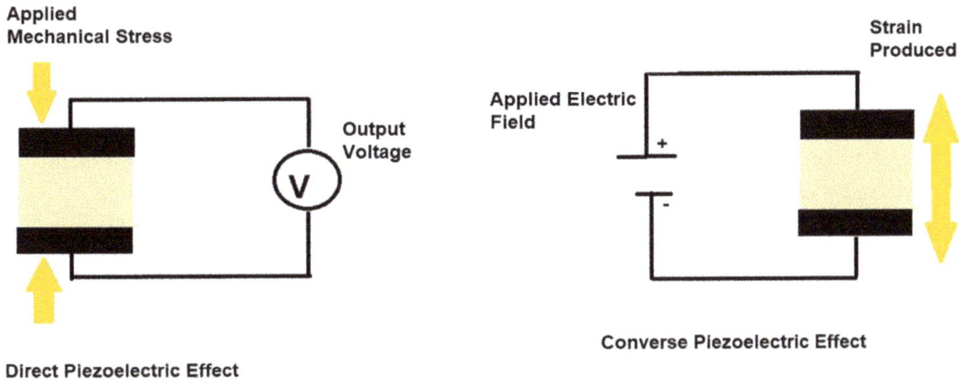

Figure 1.5. Schematic diagram of the direct and converse piezoelectric effects.

where D_i is the polarization generated along the k-axis in response to the applied stress σ_{jk}. For the converse effect S_{ij} is the strain generated in a particular orientation of the crystal by the application of the electric field E_k along the k-axis [7, 12] and d_{ijk} and d_{kij} are the piezoelectric coefficients for the direct and converse effects, respectively.

1.3.4 Pyroelectricity

Pyroelectric materials belong to a category of non-centrosymmetric polar crystals that possess coupling between the electrical polarization P and temperature T, such that a change in temperature results in a change in the electric dipole moment. In pyroelectric materials, the spontaneous polarization is not necessarily switchable using an applied electric field. These materials can be used for converting thermal energy to electric energy. Pyroelectricity is the ability of the materials to generate a temporary voltage when they are heated or cooled. As a result of the temperature difference generated across the crystal, positive and negative charge displacement takes place. Due to this relative displacement of charges, the material becomes polarized (as a molecule possesses a permanent dipole moment) and hence a potential difference is generated.

The pyroelectric coefficient (p_i) is described as the change in the spontaneous polarization with temperature [13] as

$$p_i = \frac{\partial Ps}{\partial T}, \tag{1.16}$$

where ($Cm^{-2}\ K^{-1}$) is the unit of the pyroelectric coefficient.

Alternatively, p_i is calculated using the relation

$$p_i = \frac{I}{A\frac{dT}{dt}}, \tag{1.17}$$

where I is the pyroelectric current measured during the heating cycle, A is the area of the electrode, and $\frac{dT}{dt}$ is the rate of heating. The spontaneous polarization disappears above the Curie temperature (T_c), the temperature above which a pyroelectric unit cell transforms into the centrosymmetric paraelectric phase. In the centrosymmetric paraelectric phase no dipole moment exists within the unit cell and thus it exhibits a reduced piezoelectric effect [5].

1.3.5 Ferroelectricity

The subclass of dielectric materials in which spontaneous polarization exists even in the absence of an electric field and the polarization can be reversed with the direction of the applied external field are called ferroelectric materials. Ferroelectrics are polar dielectric materials that possess at least two equilibrium orientations of the spontaneous polarization vector in the absence of an external electric field, and in which the polarization may be switched between those orientations by means of an electric field. The magnitude of the external electric field applied should be less than the dielectric breakdown of the material itself. Therefore, for the material to show ferroelectricity it must possess the following properties:

1. Spontaneous polarization.
2. The direction of polarization can be reversed by an externally applied electric field.
3. A ferroelectric hysteresis loop.
4. A ferroelectric transition temperature.

Ferroelectric materials must possess non-centrosymmetric structure. As the material is non-centrosymmetric, ferroelectricity arises due to the change in the positions of the cations and anions within the unit cell, resulting in reversible spontaneous dipole moments. This gives rise to a dipole moment equal to qd/V, where q is the electric charge on a displaced ion, d is the relative displacement, and v is the volume of the unit cell.

The dipole moment is related to the electric displacement as

$$D = \chi \chi^0 E = \chi^0 E + P, \tag{1.18}$$

where χ^0 and χ are the free space and relative susceptibilities, respectively.

Ferroelectric materials show the phase transition phenomenon. These materials undergo phase transition from ferroelectric to paraelectric above a certain temperature. During this phase transition they undergo a change in symmetry. The temperature at which the phase transition occurs is called the Curie temperature (T_c). At this temperature the symmetry of the material changes from the non-centrosymmetric to symmetric phase. Due to a change in the symmetry above the Curie temperature, spontaneous polarization will be lost and the ferroelectric material becomes paraelectric. Hence, at the Curie temperature the hysteresis curve becomes linear. In addition to the change in spontaneous polarization, temperature affects the dielectric constant of the materials.

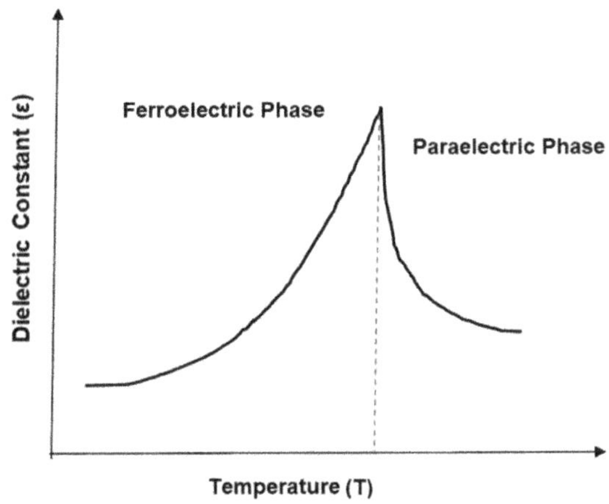

Figure 1.6. Variation of the dielectric constant with temperature.

The spontaneous polarization (P_s) decreases with an increase in temperature and vanishes at a transition temperature (T_c), while the temperature dependence of the dielectric constant (ε) shows a sharp peak at T_c as shown in figure 1.6. The Curie–Weiss temperature (T_o) coincides with the true transition temperature T_c for a ferroelectric material, which undergoes a second-order phase transition. However, in the case of a first order phase transition, the Curie–Weiss temperature T_o is several degrees lower than the transition temperature (T_c).

Some ferroelectric materials do not show a sharp peak in the temperature dependence of the dielectric constant at the transition temperature. Such materials show a broad peak at the transition temperature instead of a sharp peak. This type of dielectric behaviour is called a diffuse phase transition (DPT). It is well known that the dielectric constant/permittivity of a classical ferroelectric material above the transition temperature follows the Curie–Weiss law. However, the dependence of the dielectric permittivity of DPT ferroelectrics on a temperature above the transition temperature differs from the Curie–Weiss law over a wide temperature range. Smolenskii *et al* [14] predicted the following quadratic dependence of the dielectric permittivity on temperature:

$$\frac{1}{\varepsilon} - \frac{1}{\varepsilon_m} = \frac{(T - T_m)^2}{2\varepsilon_m \delta^2}, \tag{1.19}$$

where ε_m is the dielectric permittivity maximum, T_m is the temperature at which ε_m occurs (transition temperature), and d is the deviation factor that describes the degree of the DPT. However, Uchino *et al* [15] found that for many relaxor ferroelectric materials, the diffuseness does not exactly obey Smoleskii's [16] quadratic law and they then proposed an empirical expression to describe the diffuseness of the phase transition:

$$\frac{1}{\varepsilon} - \frac{1}{\varepsilon_m} = \frac{(T - T_m)^\gamma}{2\varepsilon_m \delta^2}. \tag{1.20}$$

In this expression γ is the degree of diffuseness and ranges from 1 to 2 for normal to relaxor ferroelectrics. The relaxor behaviour in the sample is observed due to polar domain fluctuations in the neighbourhood of the maximum permittivity and is reflected in the values of γ [17].

1.3.5.1 Hysteresis loop

The ferroelectric properties of a material are studied using a hysteresis *P–E* loop, as the ferroelectric materials exhibit hysteresis analogous to ferromagnetic materials. For these materials polarization is not a unique function of applied electric field strength. However, in dielectric materials, in general, polarization is a linear function of applied field strength, whereas it not a linear function in ferroelectric materials.

The ferroelectric specimen that is not previously subjected to an applied external electric field is said to be a virgin ferroelectric specimen. If this virgin specimen of ferroelectric material is subjected to an applied external electric field and this field is gradually increased, the polarization P increases along the curve 'ABCD' as shown in the figure 1.7. When the field gradually decreases, it is found that even when the applied electric field becomes zero ($E = 0$) there is polarization still left in the specimen. This polarization is called *remanent polarization* (P_r). In other words, the material is said to be 'spontaneously polarized'. The electric field is to be applied in the opposite direction in order to bring the polarization to zero. The required electric field to reduce the polarization to zero is called the *coercive field* (E_c).

The existence of a hysteresis loop in a material implies that the substance possesses spontaneous polarization, i.e. polarization when the applied field is zero. Extrapolation of the linear segment P_s which intersects on the P-axis at a value P_s is the maximum value of polarization attained by the specimen and is called the saturation polarization (P_s) and specimen is said to be saturated. Spontaneous polarization of a specimen depends on:

1. The dimensions of a specimen.
2. Temperature.
3. Humidity.
4. The texture of the crystal.
5. The previous history of the specimen.

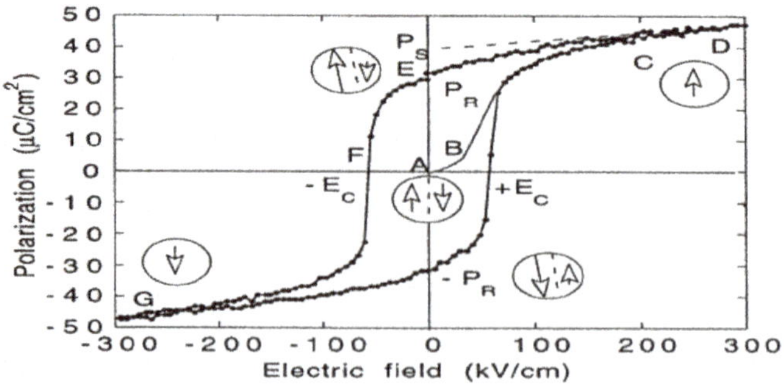

Figure 1.7. The hysteresis loop of ferroelectric materials. Reproduced with permission from [17]. Copyright 1998 IOPscience.

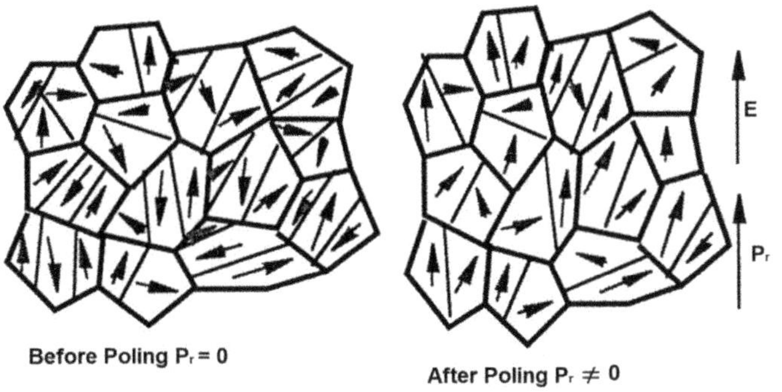

Figure 1.8. Ferroelectric domains and domain walls. Reproduced with permission from [17]. Copyright 1998 IOPscience.

1.3.5.2 Ferroelectric domain theory and domain walls

A quantitative explanation of hysteresis in ferroelectric materials is given by the domain theory of ferroelectrics. According to the domain theory, ferroelectric materials possess regions with uniform polarization called ferroelectric domains, i.e. within a domain all electric dipoles are aligned in the same direction (shown in figure 1.8).

The direction of polarization is not the same throughout the specimen. In fact, the specimen may be considered to consist of a number of domains. These domains themselves become spontaneously polarized but the direction of polarization varies from one domain to another domain. Thus, the virgin specimen may have zero polarization as a whole, i.e. the resultant polarization vector of the domains may vanish.

When an external electric field is applied, the domains in which the polarization points along the direction of the applied electric field grow at the expense of other domains in which the polarization points in the other direction. This process corresponds to the curve 'ABC' in figure 1.7.

The domains in the specimen are separated by interfaces called domain wall. The walls between domains with opposite orientation and mutually perpendicular polarizations are called the 180° and 90° walls, respectively [18]. A single domain can be obtained by domain wall motion made possible by the application of an appropriate externally applied electric field. The specific mechanism of domain wall motion is simply the small shift of the ions' positions within the unit cell resulting in the net change of orientation of the tetragonal *C*-axis. Such domain wall motion results in spontaneous polarization (shown in figure 1.8).

1.4 Classification of ferroelectrics

Ferroelectric materials can be classified on the basis of their structure and properties.

1.4.1 Classification of ferroelectrics based on crystal structure

Depending on the structure, the ferroelectric materials are classified into four major categories.

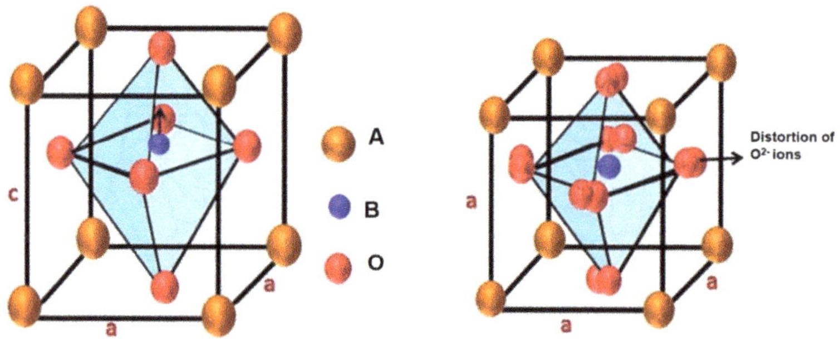

Figure 1.9. Cubic and tetragonal perovskite structures. Reproduced from [23] with permission of The Royal Society of Chemistry.

1.4.1.1 Perovskites

Perovskite is the family name of a group of materials with an archetypal ABO_3 type structure. Here the A-site is occupied by divalent cations such as Ba^{2+}, Sr^{2+}, Ca^{2+}, and Pb^{2+}, which are usually larger than the B-site cations that are tetravalent, such as Ti^{4+}, Zr^{4+}, and Sn^{4+}, and O represents the oxygen atoms. The A-site cations are surrounded by twelve anions in a cube-octahedral coordination and the B-site cations are surrounded by six anions in an octahedral coordination. Many piezoelectric (including ferroelectric) ceramics such as barium titanate ($BaTiO_3$), lead titanate ($PbTiO_3$), lead zirconate titanate (PZT), lead lanthanum zirconate titanate (PLZT), lead magnesium niobate (PMN), and potassium niobate ($KNbO_3$) have a perovskite type structure [19].

A typical ABO_3 unit cell structure of $BaTiO_3$ is illustrated in figure 1.9 and can be viewed as a closely packed cubic cell with large cations (A) at the cube corners, small highly charged cations at the body centre, and oxygen ions at the face centres. The structure is a network of corner-linked (BO_6) octahedral holes and the large cations filling the dodecahedral holes [20]. The A-site cations are in 12-fold coordination with oxygen ions and the B-site cations are in six-fold coordination. The terms A- and B-sites are used frequently in describing perovskites and perovskite-like structures. Goldschmidt [21] determined the size of the cations that can be accommodated by the perovskite structure. The A-site can contain ions with ionic radii [22] between 1.34 Å (Ca^{2+}) and 1.61 Å (Ba^{2+}) while the B-site can accommodate ions from 0.535 Å (Al^{3+}) to 0.87 Å (Ce^{4+}). Figure 1.9 shows the cubic and tetragonal perovskite structures with dipole moments. Generally, perovskite structure materials are ionic in nature.

1.4.1.2 Tungsten bronze

Tungsten bronze structure can be represented by nonstoichiometric compounds with the general formula M_xWO_3 ($0 < x < 1$) where M is an alkali metal. It is a highly coloured lustrous crystalline compound. The 'tetragonal tungsten bronze' (TTB) structure consists of interconnected corner-sharing oxygen octahedrals with three types of pseudo-symmetric open channels (three-, four-, and five-fold), which is a typical characteristic of the TTB crystal structure. It is related to the potassium

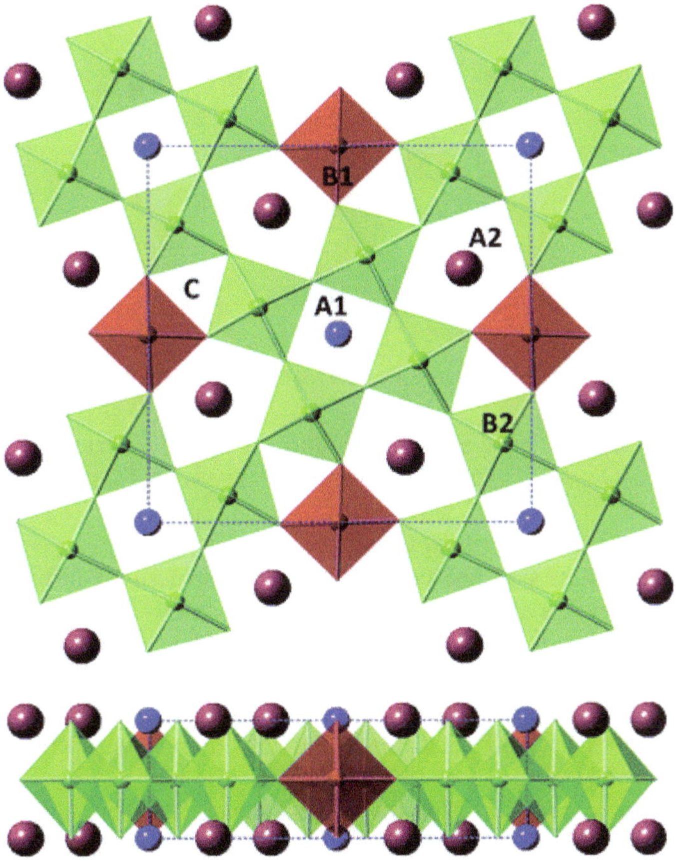

Figure 1.10. The TTB structure. Reproduced from [24] with permission of The Royal Society of Chemistry.

tungstate ($K_{0.475}WO_3$)-like structure. The general formula for the TTB ferroelectric crystal structure is given by $(A1)_2(A2)_4(C)_4(B1)_2(B2)_8O_{30}$ [24]. The TTB structure is shown in figure 1.10.

In TTB there are three open channels (the A1-, A2-, and C-sites) and the space occupied by the oxygen octahedron (B1 and B2) can accommodate a broad range of cations and anions. Hence, different properties can be tailored in TTB.

The first crystal of the tungsten bronze type structure to show useful ferroelectric properties was lead niobate ($PbNb_2O_6$) [3]. Various dopings in the TTB structure have led to the discovery of many new normal and relaxor-type ferroelectrics [25].

1.4.1.3 Pyrochlore

The general stoichiometric formula of pyrochlore structure ferroelectric material is $A_2B_2O_7$ [26], where A represents a trivalent or divalent cation and B represents a tetra- or pentavalent cation. Figure 1.11 shows the one octant part of the pyrochlore structure. Due to a broad compositional range of A- and B-site cations, pyrochlore compounds show many properties which are highly useful for anodes and cathodes

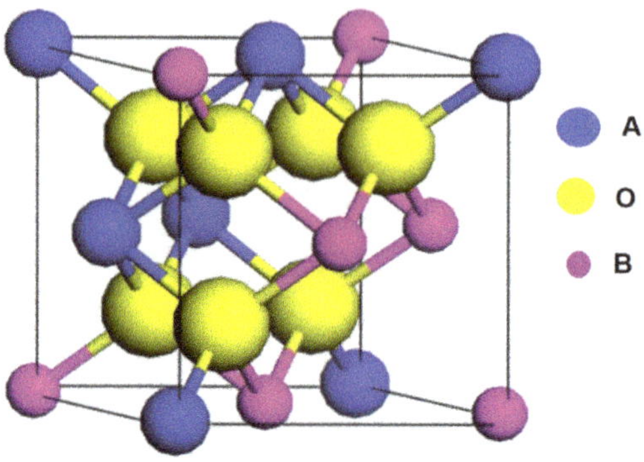

Figure 1.11. One octant part of the pyrochlore structure ($A_2B_2O_7$). [26] John Wiley and Sons, Copyright 2015.

for fuel cells and sensors, solid electrolyte catalysts and dielectrics [27]. Examples are $Ce_2Ti_2O_7$, $Sr_2Nb_2O_7$, $La_2Ti_2O_7$ (LTO), and $Nd_2Ti_2O_7$ (NTO).

1.4.1.4 Layer structured ferroelectrics

In 1949 Aurivillius first synthesized bismuth layer structured ferroelectric (BLSF) materials, hence they are also known as Aurivillius oxides [28, 29]. Bismuth layered ferroelectric materials are used mainly in storage devices and piezoelectric devices in the form of ferroelectric random access memory (FRAM), sensors, and actuators [30]. These materials possess attractive electrical properties due to their low aging and operating voltage, good retaining polarization, fatigue endurance, and high Curie temperature [31].

The Aurivillius oxides or bismuth layer structured ferroelectrics consist of perovskite-like units of TiO_2 separated by a layer of bismuth oxide. BLSFs are generally formulated as $(Bi_2O_2)^{2+}$ $(A_{m-1}B_mO_{3m+1})^{2-}$, where $(A_{m-1}B_mO_{3m+1})^{2-}$ stands for perovskite blocks interleaved with bismuth oxide layer $(Bi_2O_2)^{2+}$ along the *c*-axis. Ferroelectricity in these compounds is generally due to the cationic displacement along the polar *a*-axis and the tilting of octahedra around the *a*- and *c*-axes [32]. Some of the special properties of Aurivillius oxides are a low switching voltage, good fatigue endurance, and a lead-free chemical composition.

Bismuth based layer structured ferroelectrics are ferroelectric at room temperature and are also environmentally friendly. Their Curie temperature ranges from 300 °C to 700 °C. These materials generally have a high melting point of above 1100 °C and they also possess prospective applications in electronic equipment [31]. Due to all these properties, layered ferroelectrics have a strong potential to replace lead-based ceramics. Examples are Bi_2WO_6 ($m = 1$), $SrBi_2Ta_2O_9$ ($m = 2$), $Bi_4Ti_3O_{12}$ ($m = 3$), $MBi_4Ti_4O_{15}$ (M = Ca, Sr, Pb, Ba) ($m = 4$), $Ba_2Bi_4Ti_5O_{18}$ ($m = 5$), etc.

1.4.2 The classification of ferroelectrics based on properties

Based on their physical and electrical properties, ferroelectric materials are divided broadly into two types: (i) normal ferroelectrics and (ii) relaxor ferroelectrics.

1.4.2.1 *Normal ferroelectrics*

Normal ferroelectric materials are characterized by the following properties:

(i) A sharp first- or second-order phase transition at the Curie point (as shown in figure 1.6).

(ii) The dielectric transition temperature is independent of the frequency of the applied field.

(iii) The inverse of the dielectric constant (i.e. $1/\varepsilon$) variation with temperature follows the Curie–Weiss law, $\varepsilon = \frac{C}{T - T_0}$, where C is the Curie–Weiss constant and T_o is the Curie–Weiss temperature.

(iv) The spontaneous polarization drops to zero as the temperature of the crystal is raised to the transition temperature (T_c).

(v) A well-defined P–E loop with saturation polarization below T_c and disappear above T_c.

(vi) A large P_r (remanent polarization).

1.4.2.2 *Relaxor ferroelectrics*

Relaxor ferroelectrics are characterized by the following characteristics:

(i) A broad or diffuse phase transition.

(ii) Significant dependence of the dielectric maximum temperature (T_m) on frequency (figure 1.12).

(iii) The $1/\varepsilon'$ versus temperature curve follows the modified Curie–Weiss law (equation (1.7)).

(iv) Spontaneous polarization of the relaxor ferroelectric becomes zero at a temperature T_d (depolarization temperature) that is greater than T_m.

(v) Slim P–E loops with no sign of saturation near T_m.

(vi) Weak P_r (remanent polarization).

(vii) Exhibition of a considerable second-order piezoelectric effect by most of these materials.

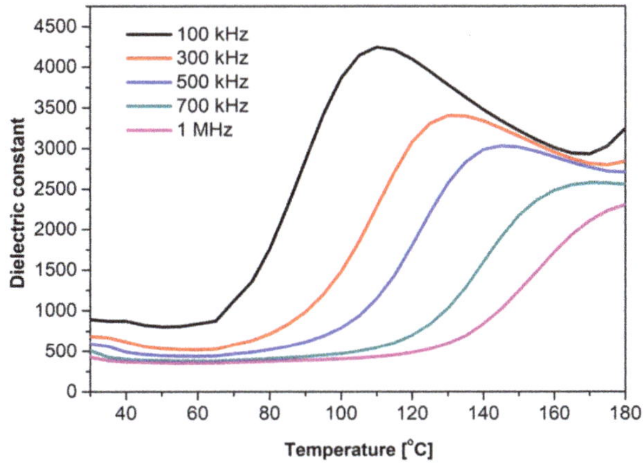

Figure 1.12. Dielectric constant of the relaxor Zr/Nb co-doped barium titanate.

1.5 Aurivillius oxides

Bismuth layer structured oxides are a family of materials first synthesized by Aurivillius in 1949 [28]. The structure of these Aurivillius phases is described as pseudo-perovskite-type layers interleaved with bismuth oxide layers. There is no particular term used to describe Aurivillius type compounds. Some of the different terminology used for these compounds are bismuth based layered compounds, perovskite based layered structures, layered bismuth oxides, bismuth layer structured ferroelectrics (BLSFs), or other similar names [29]. Most Aurivillius oxides are ferroelectric in nature and these have been the focus of research.

Layered Aurivillius oxide ferroelectrics have a general formula $(Bi_2O_2)^{2+}$ $(A_{m-1}B_mO_{3m+1})^{2-}$ [33], where m represents the number of pseudo-perovskite layers alternating between $(Bi_2O_2)^{2+}$ layers along the c-axis. The m values can range from 1 to 8. A and B are the two types of cations that enter the perovskite unit. The A-site is relatively large and is occupied by mono-, di-, or trivalent ions such as Na^+, K^+, Ca^{2+}, Sr^{2+}, Ba^{2+}, Pb^{2+}, etc, or a combination of them, and the B-site is populated by small, highly charged tetra-, penta-, or hexavalent cations such as Ti^{4+}, Nb^{5+}, Ta^{5+} of the pseudo-perovskite unit. The A-site in bismuth layer structure perovskites and ordered inter growths can accommodate cations with ionic radii ranging from 1.34 Å to 1.61 Å [32]. The B-site can accept cations of ionic radius from 0.59 Å to 0.65 Å. This is a narrower range of ionic sizes compared to perovskites. The lower limit of the ionic radii is determined by the stability of the perovskite layer and the upper limit by the mismatch between the $(Bi_2O_2)^{2+}$ and the perovskite layers. The Bi^{3+} ions occupy two distinct positions, one of which is a 12-fold coordinate site in the perovskite slabs and the other a four-fold site in the $(Bi_2O_2)^{2+}$ layers. Table 1.1 shows the family of Aurivillius compounds with a different number of perovskite layers between the Bi_2O_3 layers.

The number of perovskites between the bismuth oxide layers can be *odd, even,* or a *mixture.* The value of m influences the ferroelectric properties and structural stability of these materials. Examples of each case are Bi_2WO_6 (m = 1) and Bi_3TiNbO_9, $SrBi_2Nb_2O_9$ (m = 2) and $Bi_4Ti_3O_{12}$ (m = 3), and $CaBi_4Ti_4O_{12}$ (m = 4), and an example of a mixture of odd and even layered Aurivillius oxide is $Bi_7Ti_4NbO_{21}$. In this case, layers with 2 or 3 pseudo-perovskites are alternated with the bismuth oxide layer. Figure 1.13 shows the prototype of the Aurivillius structure as a function of the number of pseudo-perovskite layers.

The main difference between layer structured ferroelectrics and perovskite ferroelectrics is that the polarization is not produced by the displacement of the B cation within the oxygen octahedral in perovskites, while polarization is produced by relative shifting of the oxygen octahedral through the a-axis with respect to the [001] bismuth chain in BLSF. They can be divided into three main movements [35, 36]:

- Oxygen octahedral along the a-axis.
- Relative displacements in the opposite direction to the B cations of the perovskite with respect to the octahedral.
- Displacement through the a-axis of the oxygen ions of the Bi_2–O_3 layer.

Table 1.1. Aurivillius oxides with different numbers of perovskite layers between the Bi_2O_3 layer. Reproduced from [41] with the permission of AIP Publishing.

$m = 1$	$m = 2$	$m = 3$	$m = 4$	$m = 5$
$Bi_2VO_{5.5}$	$SrBiTa_2O_9$	$Bi_4Ti_3O_{12}$	$SrBi_4Ti_4O_{15}$	$Sr_2Bi_4Ti_4O_{18}$
Bi_2MoO_6	$CaBi_2Ta_2O_9$	$(Bi_{4-x}Ln_x)Ti_3O_{12}$	$BaBi_4Ti_4O_{15}$	$Ba_2Bi_4Ti_4O_{18}$
Bi_2WO_6	$BaBiTa_2O_9$	$Bi_{4-x/3}Ti_{3-x}V_xO_{12}$	$PbBi_4Ti_4O_{15}$	$Pb_2Bi_4Ti_4O_{18}$
	$PbBi2Nb_2O_9$	$Bi_{4-2x/3}Ti_{3-x}W_xO_{12}$	$CaBi_4Ti_4O_{15}$	$Bi_6Ti_3FeO_{18}$
	$SrBi_2Nb_2O_9$	$(Bi, Ln)_4(Ti, V)_3O_{12}$	$Bi_5Ti_3FeO_{15}$	$Ca_2Bi_4Ti_5O_{18}$
	$CaBi_2Nb_2O_9$	$Bi_2Sr_2Nb_2MnO_{12-x}$	$Na_{0.5}Bi_{4.5}Ti_4O_{15}$	
	$BaBi_2Nb_2O_9$	$(Bi_{4-x}Sr_2)(Ti_{3-x}Nb_x)O_{12}$	$K_{0.5}Bi_{4.5}Ti_4O_{15}$	
	$PbBi_2Nb_2O_9$	$(Bi_{4-x}Ba_x)(Ti_{3-x}Nb_x)O_{12}$		
	Bi_3TiNbO_9	$(Bi_{4-x}Pb_x)(Ti_{3-x}Nb_x)O_{12}$		
	Bi_3TiTaO_9			

$m = 1, 2$	$m = 2, 3$	$m = 3, 4$	$m = X$
Bi_2WO_5–$Bi_3Ti_{15}W_{0.5}O_9$	$Bi_3Ti_{1.5}W_{0.5}O_9$–$Bi_4Ti_3O_{12}$	$Bi_4Ti_3O_{12}$–$SrBi_4Ti_4O_{15}$	$SrTiO_3$
Bi_2WO_6–Bi_3TiNbO_9	Bi_3TiNbO_9–$Bi_4Ti_3O_{12}$ Bi_3TiTaO_9–$Bi_4Ti_3O_{12}$	$Bi_4Ti_3O_{12}$–$BaBi_4Ti_4O_{15}$	$BaTiO_3$
Bi_2WO_6–$Bi_2(Bi_{0.5}Na_{0.5})NbO_9$	$SrBi_2Nb_2O_9$–$Bi_4Ti_3O_{12}$	$Bi_4Ti_3O_{12}$–$PbBi_4Ti_4O_{15}$	$PbTiO_3$
Bi_2WO_6–Bi_3TaTiO_9	$BaBi_2Nb_2O_9$–$Bi_4Ti_3O_{12}$	$Bi_4Ti_3O_{12}(Na_{0.5}Bi_{0.5})Bi_4Ti_4O_{15}$	

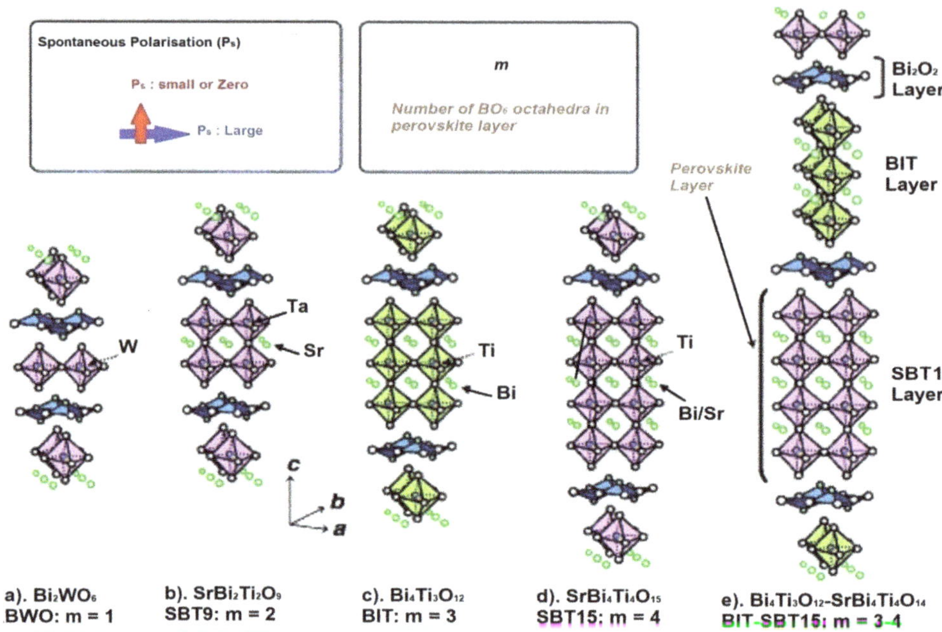

Figure 1.13. Crystal structures of Aurivillius oxide structures for m = 1, 2, 3, 4 and mixed layers. [34] (2012) Copyright. With permission of Springer.

Hence, layered ferroelectrics are ferroelectric at room temperature and are environmentally friendly. Their Curie temperature ranges from 300 °C–700 °C. BLSF materials generally have a high melting point above 1100 °C.

In addition, the peculiar crystalline structure of BLSF determines the growth habits for these materials. They also grow in anisotropic shapes, as platelets with the minor dimensions parallel to the *c*-axis. This is a major drawback for the sintering process to obtain dense bulk ceramics, but favours the grain alignment that allows the texture and the crystal orientation. Some of the special properties of BLSF are a low switching voltage, good fatigue endurance, and a lead-free chemical composition. The high coercive field in BSLF compounds makes them potential candidates for actuators applications. However, due to the high coercive field, depoling by the application of an electrical field becomes difficult, even at room temperature.

An initial study of the symmetry of Aurivillius compounds showed that the structure of these compounds is tetragonal or pseudo-tetragonal [37]. Consequently, researchers showed that the room temperature structure of most Aurivillius compounds is orthorhombic due to small distortion in the highly symmetrical tetragonal (*I4/mmm*) parent structure. Some Aurivillius compounds (Bi_2WO_6, $Ba_2Bi_4Ti_5O_{18}$) belong to the space group *B2cb* at room temperature, while some others ($Bi_2BaNb_2O_9$, $Bi_2SrNb_2O_9$) belong to the space group $A2_1am$ [38]. The number and type of shifting of perovskite layers are responsible for the low-symmetry of the ferroelectric phase and also for the high-symmetry paraelectric phase. In the family of Aurivillius oxides, the odd number layered compounds crystallize in *B2cb*, while the even number layered compounds favour $A2_1am$ symmetry. During the transition from the ferroelectric to paraelectric phase, the modification in the crystal structure of these layered compounds takes place in two ways:

i. Formation of additional chemical bonds between the cations (bismuth) in the $(Bi_2O_2)^{2+}$ layer and anions (oxygen) of the end fragments of the perovskite block at lower temperatures.
ii. Displacement of the B cation from the centre of the octahedron to the equatorial edges of the octahedron BO_6, on the basis of which the spontaneous polarization occurs in the low temperature phase [39, 40].

Currently, Aurivillius phase oxides are considered as significant ferroelectric ceramics due to their high stability, high working temperature (T_c = 550 °C–650 °C) and high operational frequency. Aurivillius oxide ceramics are used chiefly in piezoelectric resonators, as piezoelectric resonators are required to possess a very stable resonant frequency.

References

[1] Cross L E 1987 Relaxor ferroelectrics *Ferroelectrics* **76** 241–67

[2] Moulson A J and Herbert J M 1990 *Electroceramics* (New York: Chapman and Hall) p 41

[3] Valasek J 1971 The early history of ferroelectricity *Ferroelectrics* **2** 239-44

[4] Jona F and Shirane G 1962 *Ferroelectric Crystals, International Series of Monographs on Solid State Physics* (Oxford: Pergamon)

[5] Safari A, Panda R K and Janas V F 1996 Ferroelectricity: materials, characteristics and applications *Key Eng. Mater.* **122** 35–70

[6] Lencka M M, Oledzka M and Riman R E 2000 Hydrothermal synthesis of sodium and potassium bismuth titanates *Chem. Mater.* **12** 1323-30

[7] Nye J F 1985 *Physical Properties of Crystals: Their Representation by Tensors and Matrices.* (Oxford: Oxford University Press)

[8] Ahmad Z 2012 Polymer dielectric materials *Dielectric Material* (London: IntechOpen)

[9] Macdonald J R 1985 Generalizations of 'universal dielectric response' and a general distribution-of-activation-energies model for dielectric and conducting systems *J. Appl. Phys.* **58** 1971–8

[10] Sinclair D C and West A R 1989 Impedance and modulus spectroscopy of semiconducting $BaTiO_3$ showing positive temperature coefficient of resistance *J. Appl. Phys.* **66** 3850–6

[11] Duck F 2009 'The electrical expansion of quartz' by Jacques and Pierre Curie *Ultrasound* **17** 197–203

[12] Dineva P, Gross D, Müller R and Rangelov T 2014 Piezoelectric materials *Dynamic Fracture of Piezoelectric Materials* (Berlin: Springer) pp 7–32

[13] Damjanovic D 1998 Ferroelectric, dielectric and piezoelectric properties of ferroelectric thin films and ceramics *Rep. Prog. Phys.* **61** 1267

[14] Smolenskii G A, Isupov V A, Agranovskaya A I and Popov S N 1961 Ferroelectrics with diffuse phase transitions *Sov. Phys.—Sol. State* **2** 2584–94

[15] Uchino K and Nomura S 1982 Critical exponents of the dielectric constants in diffused-phase-transition crystals *Ferroelectrics* **44** 55–61

[16] Smolenskii G A E 1970 Physical phenomena in ferroelectrics with diffused phase transition *J. Phys. Soc. Jpn.* **28** 26–37

[17] Damjanovic D 1998 Ferroelectric, dielectric and piezoelectric properties of ferroelectric thin films and ceramics *Rep. Prog. Phys.* **61** 1267

[18] Stemmer S, Streiffer S K, Ernst F and Rüuhle M 1995 Atomistic structure of 90 domain walls in ferroelectric $PbTiO_3$ thin films *Philos. Mag.* A **71** 713–24

[19] Hyde B G and Andersson S 1989 *Inorganic Crystal Structures* (New York: Wiley)

[20] Merz W J 1954 Domain formation and domain wall motions in ferroelectric $BaTiO_3$ single crystals *Phys. Rev.* **95** 690

[21] Goldschmidt V M 1926 Die Gesetze der Krystallochemie *Naturwissenschaften* **14** 477–85

[22] Shannon R T and Prewitt C T 1969 Effective ionic radii in oxides and fluorides *Acta Cryst.* B **25** 925–46

[23] Kumar G R, Savariraj A D, Karthick S N, Selvam S, Balamuralitharan B, Kim H J, Viswanathan K K, Vijaykumar M and Prabakar K 2016 Phase transition kinetics and surface binding states of methylammonium lead iodide perovskite *Phys. Chem. Chem. Phys.* **18** 7284–92

[24] Gardner J and Morrison F D 2014 A-site size effect in a family of unfilled ferroelectric tetragonal tungsten bronzes: $Ba_4R_{0.67}Nb_{10}O_{30}$ (R= La, Nd, Sm, Gd, Dy and Y) *Dalton Trans.* **43** 11687–95

[25] De Araujo C P, Scott J F and Taylor G W (ed) 1996 *Ferroelectric Thin Films: Synthesis and Basic Properties* vol 10 (London: Taylor and Francis)

[26] Ren X, Zhao M, Wan C, Zheng Y and Pan W 2015 High-temperature aging of plasma sprayed quasi-eutectoid $LaYbZr_2O_7$—part I: phase evolution *J. Am. Ceram. Soc.* **98** 2829–35

[27] Megaw H 1973 Crystal structures. A working approach *Studies in Physics and Chemistry* vol (Philadelphia, PA: W B Saunders) 10

[28] Aurivillius B 1949 Mixed bismuth oxides with layer lattices *Ark. Kemi.* **1** 499

[29] Aurivillius B 1949 Mixed bismuth oxides with layer lattices I. The structure type of $CaNb_2Bi_2O_9$ *Ark. Kemi.* **1** 463–80

[30] Subbarao E C 1962 A family of ferroelectric bismuth compounds *J. Phys. Chem. Solids* **23** 665–76

[31] Newnham R E, Wolfe R W and Dorrian J F 1971 Structural basis of ferroelectricity in the bismuth titanate family *Mater. Res. Bull.* **6** 1029–39

[32] Frit B and Mercurio J P 1992 The crystal chemistry and dielectric properties of the Aurivillius family of complex bismuth oxides with perovskite-like layered structures *J. Alloys Compd.* **188** 27–35

[33] Armstrong R A and Newnham R E 1972 Bismuth titanate solid solutions *Mater. Res. Bull.* **7** 1025–34

[34] Noguchi Y and Miyayama M 2012 Defect control and properties in bismuth layer structured ferroelectric single crystals *Lead-Free Piezoelectrics* (New York: Springer) pp 405–59

[35] Rae A D, Thompson J G and Withers R L 1991 Structure refinement of commensurately modulated bismuth tungstate, Bi_2WO_6 *Acta Crystallogr.* B **47** 870–81

[36] Thompson J G, Rae A D, Withers R L and Craig D C 1991 Revised structure of Bi_3TiNbO_9 *Acta Crystallogr.* B **47** 174–80

[37] Fouskova A and Cross L E 1970 Dielectric properties of bismuth titanate *J. Appl. Phys.* **41** 2834–8

[38] Ismunandar, Kamiyama T, Hoshikawa A, Zhou Q and Kennedy B J 2005 Structural studies of five layer Aurivillius oxides: $A_2Bi_4Ti_5O_{18}$ (A = Ca, Sr, Ba and Pb) *J. Neutron Res.* **13** 183–7
[39] Subbarao E C 1962 Crystal chemistry of mixed bismuth oxides with layer-type structure *J. Am. Ceram. Soc.* **45** 166–9
[40] Newnham R E, Wolfe R W and Dorrian J F 1971 Structural basis of ferroelectricity in the bismuth titanate family *Mater. Res. Bull.* **6** 1029–39
[41] Watanabe T and Funakubo H 2006 Controlled crystal growth of layered-perovskite thin films as an approach to study their basic properties *J. Appl. Phys.* **100** 051602

Chapter 2

Odd number layered Aurivillius oxides

In this chapter, the structure-, morphology-, temperature-, and frequency-dependent properties of one-, three-, and five-layered (m = 1, 3, 5) Aurivillius phase ferroelectrics are discussed. The effects of different synthesis techniques and dopants on various properties are elaborated. The advantage of Aurivillius compounds is their compositional flexible nature which allows modification of their properties according to the chemical composition. The properties of these compounds depend closely on the type of cation occupying the A-site and B-site and the number of perovskite layers.

2.1 Single-layered oxides (m = 1) Bi_2MoO_6 and Bi_2WO_6

2.1.1 Structure

The one-layered Aurivillius oxides—Bi_2MoO_6 and Bi_2WO_6—are very high Curie temperature ferroelectrics (e.g. Bi_2WO_6 has a transition temperature of ~950 °C). Their structure below T_c has been refined into non-centrosymmetric orthorhombic space groups, showing distorted polar environments. There are three earlier reported oxide-fluoride members of this series: Bi_2NbO_5F, Bi_2TaO_5F, and Bi_2TiO_4F [1]. These compounds were initially synthesized by Aurivillius, who indexed their x-ray diffraction patterns using a body-centred tetragonal lattice. Later research [2, 3] demonstrated that these three phases were ferroelectrics with Curie temperatures of 30 °C, 10 °C, and 11 °C, respectively.

Bi_2MoO_6 and Bi_2WO_6 are chemically similar, particularly in terms of their redox chemistry and crystal structure. Their physical properties that are structure-dependent may also be similar. They both show coherent intergrowth of two closely related structures *P2₁ab* and not *B2cb*. Figure 2.1 shows a perspective drawing of the undistorted structure of Bi_2MO_6 where M = W or Mo.

Bi_2WO_6 is the simplest member (m = 1) of the BLSFs family, and has a high Curie point (above 900 °C). At room temperature, Bi_2WO_6 has the orthorhombic space group $P2_1ab$, [1, 4, 5] and not *B2cb* as reported in earlier work [6] with a = 5.457 Å, b = 5.436 Å, and c = 16.427 Å. In addition to the paraelectric–ferroelectric phase

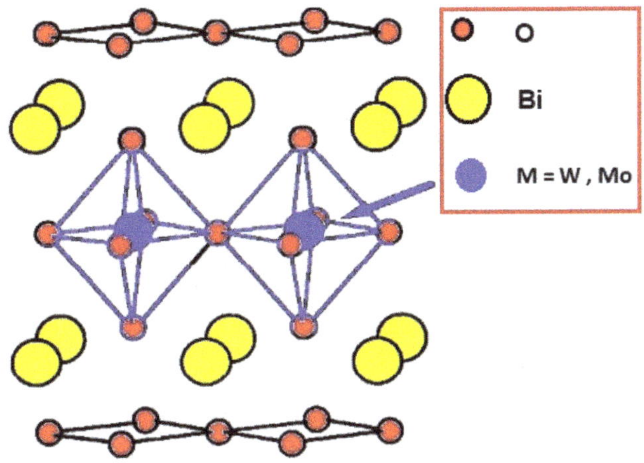

Figure 2.1. A schematic of an undistorted parent structure of single-layered Aurivillius oxides.

transition, Bi_2WO_6 ceramics have been suggested to have a second phase transition in the temperature range from 600 °C to 700 °C, accompanied by changes in lattice parameters, but their structure remains orthorhombic [7]. However, the Curie point and phase transition temperatures reported in the literature are inconsistent [6–8]. Bi_2WO_6 single crystal is considered as a high temperature piezoelectric material because of its high Curie point and piezoelectric coefficients of (d_{33} ~ 40 pC N^{-1}) [9].

2.1.2 Synthesis of Bi_2MO_6 ceramics and thin films (M = W, Mo)

Conventional solid-state synthesis of Bi_2WO_6 was reported by Juan *et al* [10]. It was observed that the amount of bismuth content does affect the phase formation and ferroelectric properties of the obtained ceramics. Bi_2WO_6 with 3.5% excess Bi_2O_3 was found to be the optimum bismuth content for single phase formation. More or less Bi content induces oxygen vacancies. An excess of Bi introduces extra Bi_2O_2 layers in the sample, leading to a well saturated hysteresis loop and higher breakdown voltage.

Ahn *et al* [11] studied the *a*-oriented and *c*-oriented polycrystalline Bi_2WO_6 thin films deposited on Pt/Ta/Glass substrate using pulsed laser deposition (PLD). The *a*- and *c*-oriented structures of BWO were confirmed from the XRD pattern. The peaks corresponding to the planes (113) and (226) were of much higher intensity in the film deposited at a higher pressure of 300 Torr as compared to those deposited at a lower pressure of 100 Torr, which confirmed the growth along the *a*-axis. For *c*-oriented Bi_2WO_6 structure, the intensity of the planes corresponding to (004), (006), and (0010) was much higher compared to the other planes. Bi_2WO_6 (BWO) thin films were epitaxially deposited onto single crystal Nb doped $SrTiO_3$ substrates via pulsed laser deposition (PLD) [12]. The deposition rate was varied from 0.01 to 0.4 nm $pulse^{-1}$, to control the growth mode of BWO thin films. The BWO thin films yielded an island growth mode during a rapid deposition rate of 0.4 nm $pulse^{-1}$. On the other hand, a layer-by-layer growth mode was achieved with a low deposition rate of 0.01 nm $pulse^{-1}$. Low deposition rate films possessed highly *c*-oriented crystallinity. It was reported that highly *a*-oriented BWO thin film possesses good ferroelectric properties with remanent polarization 14.2 μC cm^{-2} and a coercive field of 210 kV cm^{-1} due to a larger ferroelectric domain pattern as a result of smaller domain wall energy, whereas the c-oriented film of Bi_2WO_6 showed much a smaller value of remanent polarization. The *P–E* loop for the *a*-axis oriented and *c*-axis oriented structure of Bi_2WO_6 is shown in figure 2.2. According to Landau, Lifshitz, and Kittel's (LLK) scaling law, highly *a*-oriented BWO thin film would have a scaling exponent of about 0.52, which is less than the polycrystalline BWO thin film, indicating that highly *a*-oriented BWO thin film possesses a smaller domain wall energy than the polycrystalline *c*-oriented BWO thin films.

Bi_2MoO_6 is an archetypal member of the Aurivillius family of compounds with $m = 1$. It is known to possess four polymorphic phases—room temperature (γ), low temperature (γ'''), intermediate (γ''), and high temperature (γ'). At room temperature it has *Pca2₁* symmetry with alternating $(Bi_2O_2)^{2+}$ layers and perovskite-like $(MoO_4)^{2-}$ layers. At temperatures below 310 °C it exhibits polar Aurivillius phase γ'''. It shows a reversible phase transition γ' around 567 °C–607 °C with the same symmetry as $Pca2_1$ but a more distorted MoO_6 octahedral. It is reported that this phase is orthorhombic with nonpolar point group *mmm.* The high temperature irreversible phase transition γ' takes place around 710 °C–670 °C described by the $P2_1/c$ space group [13].

Mackzeka *et al* [14] reported pressure-dependent Raman scattering studies of Bi_2MoO_6 to understand the structural changes occurring in this material under hydrostatic pressure and the pressure dependence of the phonon properties. Their results indicate that Bi_2MoO_6 exhibits two structural transformations at about 2.8 and 7.0 GPa.

The cause for polymorphism in Bi_2MoO_6 as discussed by Buttrey *et al* [15] is the disproportionate in-plane modulations at lower temperatures, which tend to disappear at higher temperatures giving rise to an irreversible second order phase transition. The thermal evolution of the unequal phases is believed to be linked to

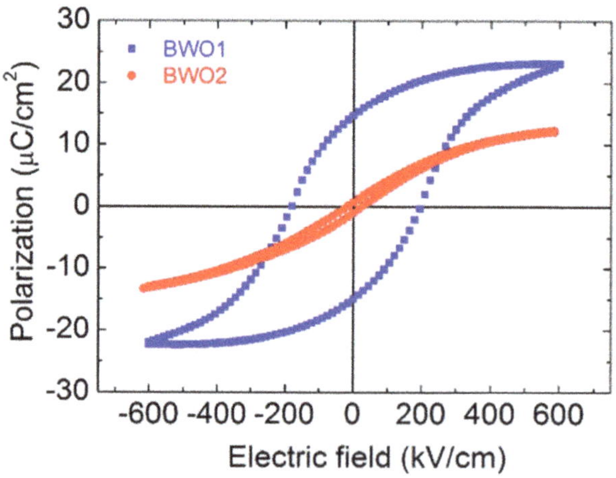

Figure 2.2. *P–E* loop for *a*-oriented and *c*-oriented structure of Bi_2WO_6. Reproduced from [12], Copyright (2017), with permission from Elsevier.

the changes in the interlayer charge transfer between two different adjacent layers $(Bi_2O_2)^{2+}$ and $(MoO_4)^{2-}$.

Polymorphism in Bi_2WO_6 and Bi_2MoO_6 Aurivillius phases was also studied by Voronkova *et al* [4] using calorimetry dielectric and electrical measurements. The results demonstrate that these compounds are similar not only in structure but also in the number and nature of their polymorphs, which were not discussed earlier. Both compounds undergo three phase transitions: from a polar orthorhombic phase into another polar orthorhombic phase ($\gamma \rightarrow \gamma'''$ ferroelectric–ferroelectric transition), then into a nonpolar orthorhombic phase ($\gamma''' \rightarrow \gamma''$ ferroelectric–paraelectric transition) and finally into a high temperature nonpolar monoclinic phase ($\gamma'' \rightarrow \gamma'$ reconstructive transition). Figure 2.3 shows the permittivity and differential scanning calorimetry (DSC) scans with temperature for BWO and BMO ceramics. In both the compounds, the ferroelectric to paraelectric transition anomalies shown by dielectric measurements corroborate well with the peaks observed in the DSC measurements.

2.2 Three-layered oxides (m = 3) $Bi_4Ti_3O_{12}$

Bismuth titanate, $Bi_4Ti_3O_{12}$ (BIT) is an example of bismuth layer structure perovskite compounds with $m = 3$ in the formula $(Bi_2O_2)^{2+}$ $(A_{m-1}B_mO_{3m+1})^{2-}$ or $(Bi_2A_{m-1}B_mO_{3m+3})$, where A = Bi and B = Ti. The crystal structure (pseudo-tetragonal symmetry) of BIT is shown in figure 2.4. Among many Aurivillius compounds, the study of bismuth titanate (BIT) has been carried out extensively by various workers [16, 17] because of its peculiar properties and possible potential applications in electronic devices. Earlier the research efforts in BIT ceramic processing were aimed at developing oriented microstructures. Recently, the focus has been shifted towards inexpensive ceramic processing which uses a mixed oxide method and pressureless sintering technique [18–20].

2.2.1 Structure

The crystal structure can be described as a stacking of layers of $(Bi_2O_2)^{2+}$ and perovskite slabs $(Bi_2Ti_3O_{10})^{2-}$ interleaved between layers of $(Bi_2O_2)^{2+}$ along the pseudo-tetragonal *c*-axis [16, 21]. In the $(Bi_2Ti_3O_{10})^{2-}$ units, Ti^{4+} ions are enclosed by oxygen octahedra, which are linked through corners forming O–Ti–O linear chains. Two of the perovskite units in the unit cell are completed by oxygen atoms that are bound to Bi in the $(Bi_2O_2)^{2+}$ layers (i.e. Bi ions in perovskite units occupy the spaces in the framework of TiO_6 octahedra). Thus the $(Bi_2Ti_3O_{10})^{2-}$ units possess a remarkable similarity to the perovskite-type structure. The height of the perovskite layer sandwiched between the $(Bi_2O_2)^{2+}$ layers in BIT is equal to six Ti–O distances or approximately three ABO_3 perovskite units. The Bi^{3+} ions occupy two distinct positions, one of which is a 12-fold coordinated site in perovskite slabs and the other a four-fold site in the $(Bi_2O_2)^{2+}$ layers. BIT, unlike the other Aurivillius phases, is monoclinic with space group *B1a1* [22] in the ferroelectric phase. The dimensions of the monoclinic BIT unit cell are a = 5.450 Å, b = 5.4059 Å, and c = 32.832 Å [22]. The ferroelectric Aurivillius phases, including BIT, are usually described on the basis of a doubled orthorhombic diagonal cell, which transforms

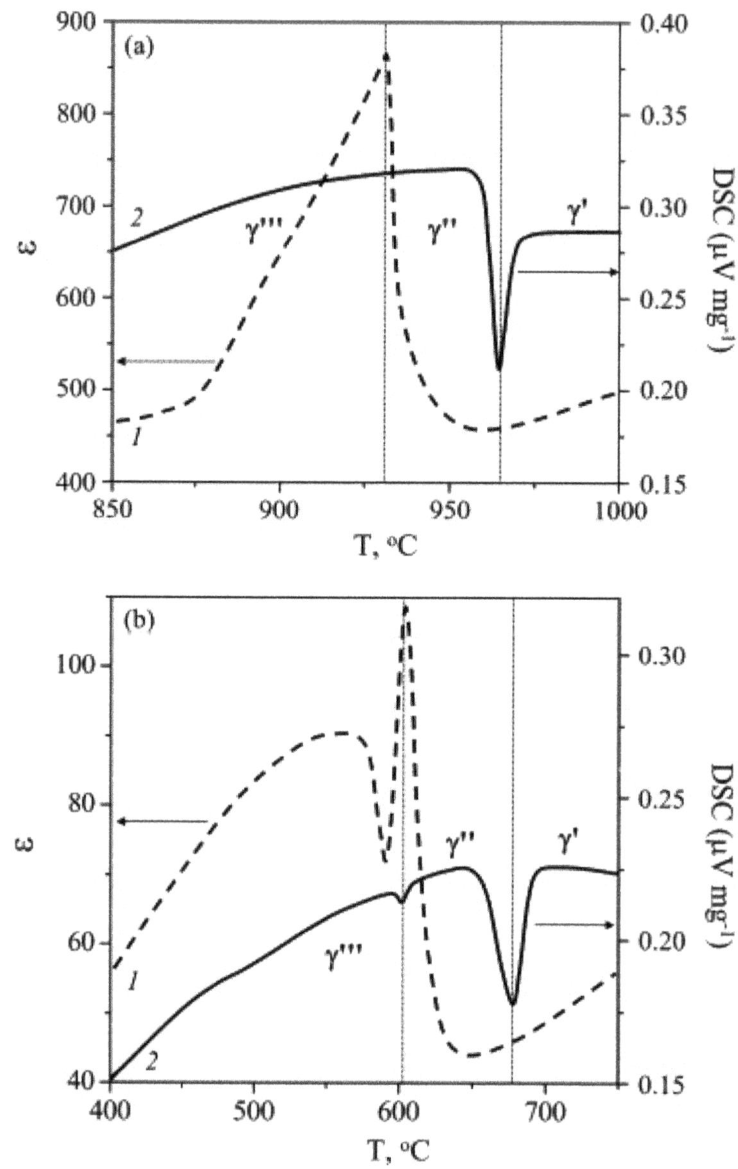

Figure 2.3. Variation of permittivity and DSC scans with temperature for (a) BWO and (b) BMO. Reproduced from [4], Copyright (2009), with permission from Elsevier.

the tetragonal *I4/mmm* prototype structure to an orthorhombic *Fmmm* underlying nonpolar parent structure [23].

BIT bears a special interest in the Aurivillius family of compounds because of its ferroelectric properties, which has switching applications with low leakage, fatigue, and high remanent polarization [25]. It is an unusual ferroelectric material, as it has two components of polarization vector out of which one can be switched at low

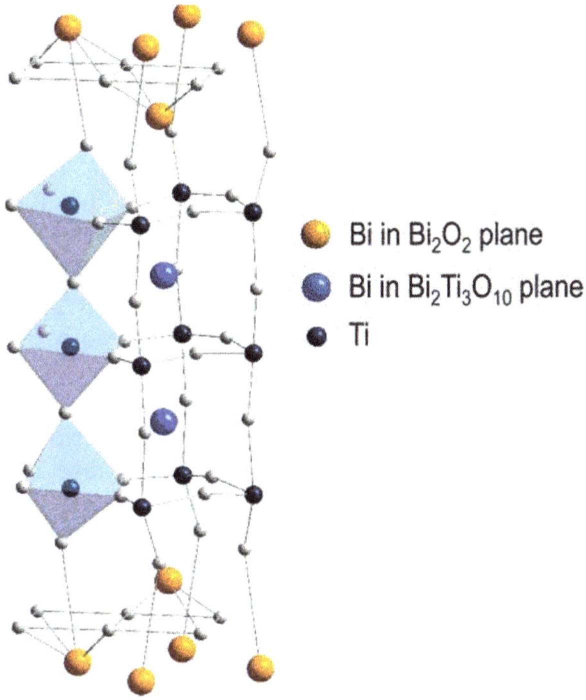

Figure 2.4. Crystal structure of bismuth titanate $Bi_4Ti_3O_{12}$. Reprinted from [24], with permission of AIP Publishing.

fields along the *c*-axis and the other at high fields along the *ab*-axis [10]. Due to these properties, this material has been identified as a promising candidate to be used for nonvolatile memory and dynamic memory of computers. Among the many ferroelectric materials, bismuth titanate ($Bi_4Ti_3O_{12}$) and its cationically substituted derivatives are taken into consideration [26]. BIT has a high Curie temperature (T_c). It shows ferroelectric to paraelectric phase transition at about 675 °C [27–29]. Several structural studies at room temperature and at higher temperatures are available in the literature, explaining the crystal structure of both ferroelectric and paraelectric phases [22, 27, 28, 30]. The early powder XRD studies on BIT indicated an orthorhombic lattice with space group *Fmmm* [21, 31, 32]. Later the detailed crystal structure of the ferroelectric phase of BIT showed a similar orthorhombic lattice but the space group was changed to *B2cb* [28, 30]. Further, the crystal structure was re-determined by Rae *et al* [22] and Shimakawa *et al* [33] and they reported a monoclinic *Pc* lattice. However, the monoclinic distortion in the lattice is reported to be too small to differentiate these two lattice types. Kim *et al* [34] studied this compound using combined neutron and x-ray diffraction of the polycrystalline sample and supported the monoclinic lattice. Recently, the Rietveld refinement of the observed low temperature powder neutron diffraction data revealed the retention of the orthorhombic lattice with space group *B2cb* without any significant change in the atomic arrangement [35].

2.2.2 Synthesis methods

The performance of materials is related closely to the way they are processed. To determine the microstructure, and the electrical and optical properties of bismuth titanate ceramics and thin films, the synthesis method plays a vital role in optimizing the performance of the material.

BIT has been prepared by a number of methods to reduce the sintering temperature and hence reduce the grain growth. These methods include conventional solid-state reaction [36], sol–gel [37], hydrothermal [38], co-precipitation [39], and metal–organic decomposition [40] methods. The solid-state method has a limitation of high sintering temperature causing Bi^{3+} ion volatilization which occurs at temperatures above 1130 °C [41]. Further, the solid-state technique produces a powder with large particle size and broad of particle size distribution. The sol–gel technique is used for producing nanoparticles, but it requires long processing periods and repeated heat treatments. The hydrothermal technique has the advantage of producing a fine particle size but the stoichiometry of the products is dependent on the kinematics and thermodynamics of the elements during the reaction [41]. The co-precipitation technique has the possibility of the retention of impurities and synthesis in an aqueous environment and causes water to be incorporated into the powder, thus causing deterioration in electrical properties.

A bismuth titanate sample obtained through the conventional solid-state route possessed orthorhombic symmetry with $a = 5.411$ Å, $b = 5.448$ Å, and $c = 32.83$ Å at room temperature (25 °C). A high temperature x-ray study revealed a symmetry change to tetragonal at 675 °C and suggested that the orthorhombic *b*-axis is probably the polar axis in BIT by comparing the lattice parameter data with those of the perovskite-type ferroelectrics. It was also reported that the multiple ion substitution of Bi^{3+} and Ti^{4+} in BIT by a combination of divalent and pentavalent ions such as Ba–Nb, Pb–Nb, and Sr–Nb leads to a steep decrease in the Curie temperature of BIT [21].

The behaviour of oxygen vacancies in the BIT ceramics prepared through the conventional solid-state reaction method was investigated by Li *et al* [42] using dielectric measurements. The Cole–Cole fitting to the dielectric loss peak reveals a strong correlation between oxygen vacancies, which was considered to exist commonly in ferroelectrics. Therefore, the migration of oxygen vacancies in ferroelectric materials would demonstrate a collective behaviour due to this strong correlation. Furthermore, this correlation was reported to be greatly concentration-dominated and has an apparent effect on activation energy.

The grain growth behaviour of BIT ceramics synthesized by the solid-state route under different sintering conditions was reported by Watcharapasorn *et al* [43]. It was found that the sintering of ceramics for a longer duration could render a material with a more isotropic microstructure and with reduced preferred orientation. Large values of remanent polarization $P_r = 24\ \mu C\ cm^{-2}$ and a low coercive field $E_c = 11\ kV\ cm^{-1}$ were obtained for BIT ceramics with the density of 98% after low temperature sintering at 850 °C for the powder derived from a mechanical activation technique [44]. Stojanovic *et al* [45, 46] reported that the BIT powder can be synthesized directly using high impact milling for about 3 to 12 h and then sintered at 1000 °C for 2 h. Han *et al* [47] stated that the formation of the BIT phase is highly dependent on the processing parameters, particularly the impact energy or milling intensity. Zdujic *et al* [48] reported that a mixture of α-Bi_2O_3 transformed to $Bi_2O_2CO_3$ at a milling intensity of ~0.49 W g^{-1}, which in turn was converted directly into a nanocrystalline BIT phase when the intensity was increased to ~ 2.68 W g^{-1}.

Highly stable and homogeneous BIT powder was produced by the sol–gel synthesis technique at calcination temperature as low as 550 °C, which is relatively low for the wet chemical technique [49, 50]. Pookmanee *et al* [51] reported the phase structure and particle morphology of hydrothermally synthesized bismuth titanate powder which was significantly influenced by different mineralizer KOH contents and molar ratios of Bi/Ti [52]. Also, Xie *et al* [53] reported that the concentration of KOH, reaction time, and temperature had a significant effect on the phase composition and morphology of the resultant single crystals. Kan *et al* [54] reported that well dispersed particles of about 10 nm began to form a BIT phase at 470 °C. The complete phase formation was achieved after a heat treatment at 550 °C for 30 min. It was finally sintered at 750 °C for 1 h to achieve a sample with a high density of 97.2%.

Polycrystalline powders of bismuth titanate were prepared using a solution combustion technique by Subohi *et al* [55, 56]. This process involves self-sustained reaction in a homogeneous solution of different oxidizers (e.g. metal nitrates) and fuels. Depending on the type of precursors as well as the conditions used for process organization, solution combustion synthesis (SCS) may occur as either volume or self-propagating high temperature synthesis modes. This process not only yields nano-sized oxide materials but also allows uniform dropping of trace amounts of impurity ions in a single step. Some of the salient advantages of the SCS process are as follows:

(i) It is an easy and fast process that uses relatively simple equipment.
(ii) The composition, structure, homogeneity, and stoichiometry of the products can be controlled precisely.
(iii) The formation of high purity products is ensured.
(iv) The high exothermicity of the metal nitrate–fuel reaction permits incorporation of the desired quantity of impurity ions or dopants in the oxide hosts, to prepare industrially useful materials.
(v) The stabilization of metastable phases (γ-Fe_2O_3, t-ZrO_2, anatase TiO_2, etc) is possible.
(vi) The formation of products of virtually any size (microns to nanometres) and shape (spherical to hexagonal) can be achieved.
(vii) It is a cost effective method compared to conventional ceramic methods and is easy to scale up.

Fuel plays an important role in combustion synthesis. The fuel forms complexes in metal ions facilitating homogeneous mixing of cations in a solution. During combustion it breaks down into components, which in turn decompose to produce combustible gases such as HNCO and NH_3, which ignites with NO_x [57]. There are several fuels available for combustion synthesis, such as urea, glycine, hexamethyl-tetramine, carbohydrates, oxalic acid, citric acid, etc. However, compounds containing N–N bonds are found to assist the combustion better.

The maximum reaction temperature generated in this process depends on the fuel to oxidizer ratio, initial furnace temperature, nature of the fuel, and quantity of initial precursors and, among these, more strongly on the nature of the fuel [58, 59]. A good fuel should have excellent complexing ability and produce sufficient enthalpy and a high volume of gases. Fuels differ in their reducing powers and the number of moles of gases they generate. Hence, they have their own effects on the nature of the final product and its properties [60]. Therefore, it is imperative to study the effect of the type of fuel on the microstructure and properties of the processed ceramics.

Based on the above study, three samples of bismuth titanate ceramics were prepared using bismuth nitrate and titanium isopropoxide as cation precursors and three different fuels, namely glycine, urea, and dextrose. Each of the three fuels differs in its reducing power and generates a different number of moles of gases [61]. The reaction mechanism for each fuel is given below.

For dextrose as fuel:

$$4Bi(NO_3)_3.\ 5H_2O + 3TiC_{12}H_{28}O_4 + 2.5\,C_6H_{12}O_6 \rightarrow Bi_4Ti_3O_{12(s)} + 27CO_{2(g)} + 6N_{2(g)} + 17H_2O_{(g)} + 60H_{2(g)} + 24C_{(s)}. \tag{2.1}$$

For glycine as fuel:

$$4Bi(NO_3)_3.\ 5H_2O + 3TiC_{12}H_{28}O_4 + 6C_6H_5O_2N \rightarrow Bi_4Ti_3O_{12(s)} + 25CO_{2(g)} + 9N_{2(g)} + 18H_2O_{(g)} + 59H_{2(g)} + 23C_{(s)}. \tag{2.2}$$

For urea as fuel:

$$4Bi(NO_3)_3.\ 5H_2O + 3TiC_{12}H_{28}O_4 + 10CH_4N_2O \rightarrow Bi_4Ti_3O_{12(s)} + 24CO_{2(g)} + 16N_{2(g)} + 18H_2O_{(g)} + 64H_{2(g)} + 22C_{(s)}. \tag{2.3}$$

The effects of the addition of these fuels on the combustion behaviour, resulting powder characteristics, and the dielectric and ferroelectric properties were studied [61]. Different fuels have different complexing power, which affects the phase formation and morphology of the synthesized powder. The chemical energy released during the exothermic reaction between the metal nitrates and various fuels is different. As per the propellant chemistry, the ratio and oxidant are fixed in such a way that the net reducing valency of the fuel equals the net valency of the oxidant. The reducing valencies of the fuels are given in table 2.1.

It was observed that during the combustion reaction the samples with stoichiometric urea and glycine as fuel produce a vigorous reaction with a high flame, thus producing voluminous amounts of a yellowish-grey powder, whereas for the sample with dextrose, the combustion reaction is flameless in nature and produces a dry, loose, and black powder. This shift in colour from yellow to black may be due to the presence of carbon residues in the processed powder. Since the carbon is oxidized and evaporates during combustion, the residual carbon content in the combustion

Table 2.1. The reducing valency of the different fuels used and the Lotgering factor as obtained from XRD patterns.

Fuel	Reducing valency	Lotgering factor (α_c)
Urea	+6	57%
Glycine	+9	60%
Dextrose	+24	38%

synthesized powder is inversely proportional to the flame temperature generated by the fuel [59]. The powder then turns white after calcination at 800 °C for 3 h as the carbon residues are removed.

X-ray diffraction analysis at three different stages of sample preparation, (i) after combustion, (ii) after calcination, and (iii) after sintering, helps in understanding the phase formation in bismuth titanate ceramics. After combustion all three samples were still amorphous in nature. In the samples with glycine and dextrose as the fuel, the presence of secondary phases was more prominent, whereas in the case of the sample with urea as the fuel a smaller number of secondary phases were present. This may be due to the lower flame temperature in case of dextrose and glycine and the higher flame temperature in the sample with urea as the fuel. Completely crystalline samples with no secondary phases were obtained only after calcination at 800 °C for 3 h. The diffraction patterns of sintered ceramics show an increase in the intensity of the peaks corresponding to the planes (00*l*). This indicates that as the ceramics are sintered at higher temperatures, a greater number of planes become oriented parallel to the crystallographic plane (00*l*). This may be attributed to the fact that the growth rate of the ceramic is higher in the direction perpendicular to the bismuth layer (the crystallographic *c*-axis) than the direction parallel to the bismuth layer (along the crystallographic *a*- or *b*-axis) [62]. The reason is that the compacted particles with the *c*-axis oriented parallel to the pressing direction tend to grow and contribute to a high degree of preferred orientation (00*l*) [43]. The XRD patterns of the as-prepared samples were indexed based on those given in JCPDS data card file no.72–1019. The degree of *c*-axis orientation is determined by the Lotgering factor (α_c) given by [61]

$$\alpha_c = \frac{I(0014)}{[I(0014) + I(117)]}. \tag{2.4}$$

The (α_c) calculated for samples with different fuel are listed in table 2.1, which indicates clearly that growth along the *c*-axis is less in the sample with dextrose as the fuel, as the value of α_c is lowest in the case of the sample with dextrose as the fuel. It was observed that the dielectric loss and ferroelectric to paraelectric transition temperature T_c = 650 °C remains unaffected by the variation in fuel, as these are structure-dependent parameters and all the three samples possess an orthorhombic structure below T_c. The sample with dextrose as the fuel shows a higher dielectric constant and higher remanent polarization compared to the samples with urea and glycine as the fuel due to a smaller grain size and less c-axis grain growth, respectively

The analysis of SEM micrographs of the calcined powders reveals the formation of large agglomerates due to the prevalence of high temperatures and the evolution of a smaller number of moles of gases when urea is used as the fuel. Porous powders with a foamy aspect were seen with glycine and dextrose as the fuel due to the lower flame temperatures and the evolution of large amounts of gases. The images of the sintered ceramics show complete grain formation with different porosity. It can be understood clearly that the grain size is smaller in the case of dextrose-based BIT (around 5 μm) than that for glycine-based BIT (about 10 μm) and urea-based BIT (>10 μm). The dependence of the dielectric constant on temperature at different frequencies for bismuth titanate ceramics obtained by the solution combustion route with dextrose as fuel is shown in figure 2.5(a). It is noted that the relative permittivity undergoes low frequency dispersion for all temperatures due to the presence of the conduction mechanism in the sample. Here, considering the charge carriers in the polarization mechanisms, the relative permittivity behaves like a conductive component. The anomalous peak observed at 550 °C was also reported earlier

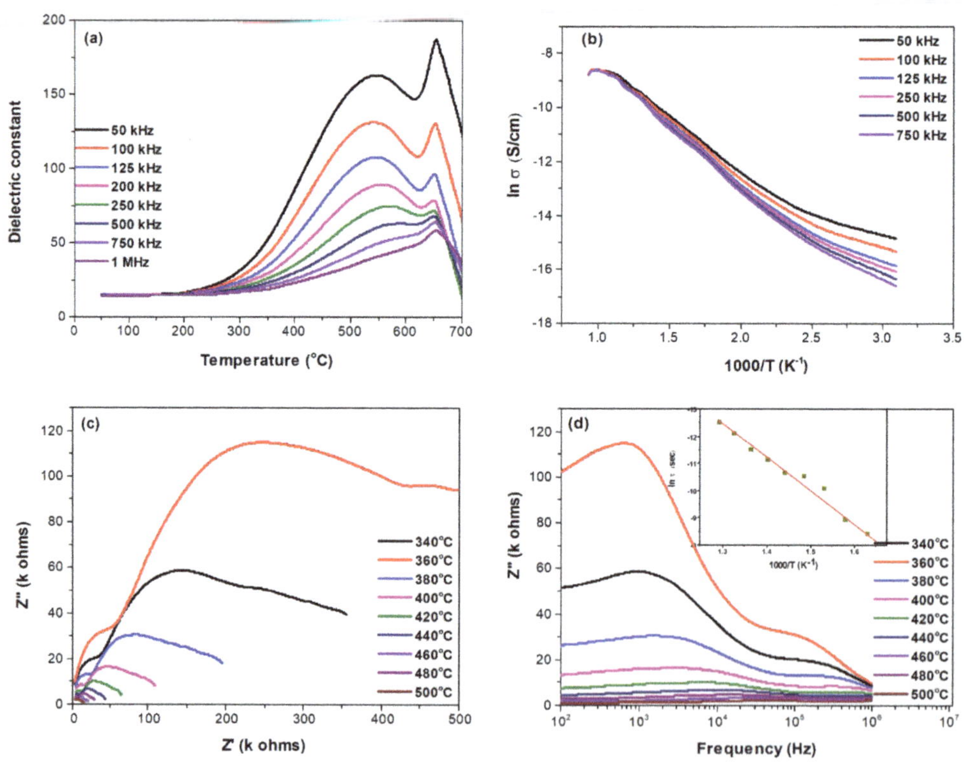

Figure 2.5. (a) Variation of ε with temperature at different frequencies. (b) Arrhenius plots, (c) Nyquist plots, and (d) variation of Z' with frequency at different temperatures for BIT ceramics obtained through the solution combustion route with dextrose as the fuel. The inset shows the relaxation time with the inverse of temperature.

and is attributed to the space charge generated at the surface of the ceramics and to the ion-jump relaxations [63]. This anomalous peak is more prominent at lower frequencies and is absent at higher frequencies. This can be explained by the fact that defects, i.e. oxygen vacancies, present in the sample create a space charge whose polarization responds to the applied electric field. At low frequencies, these charges have enough time to move longer distances in the sample creating larger electronic polarization and therefore a high value of dielectric constant is measured. However, at higher frequencies the space charge no longer follows the field, leading to lower values of the dielectric constant. With increasing temperature, the effect of charge polarization is increased, thus giving higher values of relative permittivity. At a certain temperature the space charge is either trapped at the electrode ferroelectric interface or neutralized. The trapped space charge reduces the electric field felt in the bulk of the sample resulting in a poor polarization response and hence lowers the value of the dielectric constant. As the temperature is further increased, more dipoles are able to switch leading to an increase in the value of the dielectric constant.

Figure 2.5(b) shows that the conductivity in the BIT ceramics prepared with dextrose as the fuel increases with increasing temperature. During the preparation of this compound it was sintered at a high temperature. Thus there is a possibility for the loss of oxygen, thus creating vacancies as per thermal equilibrium conditions. When the samples are cooled to room temperature, oxygen re-enters the lattice, but may not compensate totally for the loss of oxygen during sintering. The entry of oxygen is compensated only on the surface while the interior of the sample has an oxygen vacancy. The oxygen loss is denoted as per the Kroger–Vink reaction [64]:

$$O_o^x \leftrightarrow 1/2O_2'' + V_o'' + 2e', \tag{2.5}$$

where e' represents the electron and $V_o^{'}$ represents the oxygen vacancy. Therefore, the conductivity of the samples can be associated with the mobility of oxygen vacancies. The activation energy at different frequencies is calculated with these plots using the relation [65]

$$\sigma = A\exp\left(-\frac{E_a}{kT}\right), \tag{2.6}$$

where σ represents the conductivity of the sample, A is the pre-exponential constant, E_a is the activation energy, k is the Boltzmann constant, and T is the absolute temperature.

Nyquist plots of BIT synthesized by solution combustion using dextrose as the fuel are presented in figure 2.5(c). The semicircles are well resolved—each of them corresponds to the impedance of different microstructural regions of the ceramic. The semicircles have their centres located below the real impedance axis (Z'), indicating the presence of two or more relaxing charged defects and therefore indicating a Maxwell–Wagner (M-W) relaxation behaviour [66]. The maximum of each semicircle corresponds to the relaxation frequency and the radius of the semicircle gives the resistance [67]. The radius of the semicircles decreases with an increase in the temperature, indicating a decrease in relaxation time. Complex impedance behaviour can be described by a series of three parallel R–C circuit elements that correspond to the dielectric behaviour of the grain bulk, the grain boundary, and the electrode film, respectively. These are characterized by the parameters (R_g, C_g), (R_{gb}, C_{gb}), and (R_e, C_e), where R represents the resistance, and the capacitance C represents the polarization of these regions. The data were fitted using an equivalent circuit with electrochemical impedance spectroscopy (EIS) software. For quantitative analysis of the spectra, a simplified equivalent circuit of resistors in series—each shunted by a constant phase element (CPE)—was used. The CPE represents the inhomogeneous distribution of grains and their sizes. The grain resistance (R_g) and grain boundary resistance (R_{gb}) can be obtained from the intercept on the real (Z') impedance axis. Also, the capacitances (C_g) and (C_{gb}) can be obtained from the frequency corresponding to the highest point in each semicircle, using the relation $2f_{\max}RC = 1$, where $f_{\max}$ is the relaxation frequency. It is noted that $R_g << R_{gb}$ indicates that the grain resistance is the dominant factor for the overall electrical behaviour of the as-prepared ceramic in the temperature range 340 °C to 500 °C. The variation of the imaginary part of the impedance (Z'') with frequency at different temperatures for BIT shown in figure 2.5(d) indicates that the Z'' value increases initially, reaches the maximum value (Z_p, the peak value) and then decreases with an increase in frequency for all temperatures. It is noted that the Z_p value decreases with increasing temperature, denoting the increase in capacitance and decrease in resistance of the material, indicating an increase in conductivity. In addition, as the temperature increases the peak tends to broaden and the frequency corresponding to Z_p shifts towards a higher frequency. This broadening of the peak indicates multiple relaxations in the material. These multiple relaxations may be due to fluctuations in the structure because of the random occupation of bismuth at certain sites.

The binding energy of the electrons, that is the energy required to move the electrons from one site to another at different temperatures, is calculated using the relation

$$\beta = \frac{6K_BT}{W_m}, \tag{2.7}$$

where $\beta = (1-s)$, W_m is the binding energy in electronvolts, and s is the slope obtained from the σ versus frequency curves. It is observed that the binding energy decreases with an increase in temperature, i.e. as the temperature is increased, less energy is required to move the electrons from one position to another, thus supporting the fact that conductivity increases with an increase in temperature.

The minimum distance moved by the electron ($R_{\min}$) is calculated for different temperatures from

$$R_{\min} = \frac{2e^2}{\pi\varepsilon\varepsilon_0 W_m}. \tag{2.8}$$

It is noted that $R_{\min}$ increases with an increase in temperature, as the temperature increases conductivity due to the dominant long range transport of charge carriers.

Based on the correlated barrier hopping (CBH) model, the ac conductivity data can be used to evaluate the density of states at the Fermi level $N(E_f)$ using the [68]

$$\sigma_{ac} = \frac{\pi}{3} e^2 \omega k_B T \{N(E_f)\}^2 \alpha^{-5} \left\{ \ln\left(\frac{f_0}{\omega}\right) \right\}^4, \tag{2.9}$$

where f_o is the photon frequency and α is the localized wave function, assuming $f_o = 10^{13}$ Hz, and the polarizability $\alpha = 10^{10}$ m^{-1} at various operating frequencies and temperatures [69]. It was observed that $N(E_f)$ decreases with an increase in frequency for all temperatures and merges above 50 kHz. It is clear that $N(E_f)$ increases with temperature at lower frequencies and is independent of temperature at higher frequencies.

Cummins *et al* [31] reported that in single crystals, the polarization vector in BIT is inclined at a small angle (~ 4.5°) to the *a–b* plane which leads to two polarization moments: one along the *a*-axis of orthorhombic structure, $P_s = 50\ \mu C\ cm^{-2}$, which is twice as large as that of $BaTiO_3$ and the other along the *c*-axis, $P_s = 4\ \mu C\ cm^{-2}$. The crystal structures of Bi_2WO_6 and $Bi_4Ti_3O_{12}$ have been refined from x-ray and neutron diffraction data by Newnham *et al* [41] elucidating the distortions responsible for ferroelectricity. They have also reported that Bi_2WO_6 is orthorhombic, space group *B2cb*, $Bi_4Ti_3O_{12}$ is monoclinic, space group *Pc*, but very nearly orthorhombic, space group *B2cb*. They observed extremely short bond lengths (1.7 Å) at the TiO_6 octahedra in BIT and even in $BaTiO_3$ and $PbTiO_3$; the shortest Ti–O distances are 1.87 Å and 1.79 Å, respectively. They concluded that sizeable distortions are to be expected in the bismuth titanate family because of their large polarization and high Curie temperature. They also reported that at the phase transition a strong bond is formed between the Bi ion of the Bi–O layer and the apex oxygen of the perovskite layer. They also reported that the bond direction lies in (*l*00) perpendicular to the polar axis and gives rise to anti-parallel displacement along the *b*- and *c*-axes.

Newnham *et al* [41] identified the major contributors to the spontaneous polarization in bismuth layer structure perovskites as the octahedral cations (Ti^{4+}) moving towards the octahedral edges. Thompson *et al* [22] reported that the main cause of ferroelectricity in $Bi_4Ti_3O_{12}$ is the *a*-axis displacement of Bi^{3+} atoms in the perovskite A-site with respect to the chains of TiO_6 octahedra.

Maffei *et al* [26] carried out dielectric and ferroelectric studies on *c*-axis preferred oriented BIT film. They reported a dielectric constant of 150, a loss tangent of 0.01 at 100 kHz, a remanent polarization of approximately 7 $\mu C\ cm^{-2}$, and a coercive field of 20 kV cm^{-1}. They attributed this high coercive field to the presence of a minor *a*-axis polarization component, as evident from the XRD data. The synthesis, dielectric, electrical, and piezoelectric properties of pure, Fe, and Nb doped BIT at the B-site, prepared through the solid-state reaction method in different atmospheres, were reported by Shulman *et al* [20, 57].

The ferroelectric and dielectric properties of pure BIT ceramics prepared using the conventional solid-state reaction method were investigated by Li *et al* [70] after different processes of oxygen treatment. They highlighted that both the remanent polarization and fatigue resistance are enhanced with the decrease of oxygen vacancies inside the ceramics. It was also reported that the Cole–Cole fitting of the dielectric loss peak reveals a strong correlation among oxygen vacancies, which exist in oxide ferroelectrics. Therefore, the migration of oxygen vacancies in ferroelectric materials would demonstrate a collective behaviour due to this strong correlation.

Pyroelectricity was observed in bismuth titanate ceramics and thin films. The variation of the pyroelectric coefficient with temperature for BIT ceramics sintered at 850 °C was studied [71] and a broad peak around 100 °C was reported. The pyroelectric coefficient for the poled sample was found to be higher than for the unpoled sample. However, a shift in the peak to 175 °C was reported. This indicates a rather strong influence of processing and measurement conditions on the pyroelectric properties of BIT. Higher values of pyroelectric coefficient were reported in the *c*-oriented BIT thin films shown in figure 2.6 [72]. The pyroelectric response from a single dot to a dark-light pulse was examined using a 150 W incandescent lamp. Incident light passed through a square aperture in the chopper and was focused by a convergent lens. As the light was switched on, a peak response was detected. This response was assumed to result from three mechanisms [73]: (i) pyroelectric, from the rapid conversion of absorbed optical radiation to heat, (ii) piezoelectric, from the relaxation of an initial state of stress produced by the absorbed energy, and (iii) photovoltaic, from the optical production of carriers and their net transport by internal fields. The response decays to base levels in about 20 s after the light was switched off. Films with higher values of the degree of orientation (α_{00l}) give a higher response. The magnitude of these peaks was in the range of (3.2–4.6 mV). Dynamic pyroelectric responses to illumination from the 150 W lamp modulated by a chopper

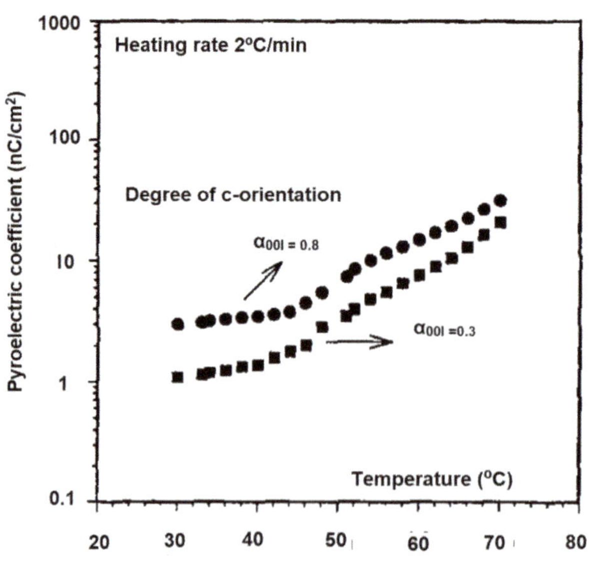

Figure 2.6. Temperature dependence of the pyroelectric response of bismuth titanate films. [72] (1997) Copyright. With permission of Springer.

and focused by a convergent lens were recorded for films with $\alpha_{00l} = 0.25$, 0.6, and 0.8. It confirmed the fact that films with a higher degree of *c*-orientation produce higher pyroelectric responses. This behaviour shows a small change in *z*-polarization with temperature for the optimized *c*-oriented films, as confirmed by hysteresis loop measurements.

2.2.3 The effect of A-site and B-site doping in $Bi_4Ti_3O_{12}$

Despite having aroused considerable interest regarding its potential, $Bi_4Ti_3O_{12}$ (BIT) possesses high dielectric loss and is difficult to pole, and shows small remanent polarization (P_r). These inferior properties are believed to be due to the instability in the oxidation state of the Ti^{4+} ions [74] and the defects caused due to the volatile property of bismuth during the sintering process accompanied by oxygen vacancies [75]. Therefore, modification of the basic compound by ion substitution either at the A-site or B-site in BIT was reported to tailor its properties. In addition, the valence state of the dopant, its ionic radius, and the position of the dopant will affect the properties of the basic compound.

The effect of the incorporation of smaller trivalent cations into the bismuth sites on the crystal and electronic structures of ferroelectric bismuth titanate has been investigated by Hur *et al* [76]. From x-ray diffraction analyses on $Bi_{3.25}A_{0.75}Ti_3O_{12}$ (A = Bi, La, Gd, and Y), they found that the cation substitution gives rise to the contraction of the volume of the unit cell and decreases the structural distortion of bismuth titanate. Their x-ray absorption spectroscopy results reveal that the local structural change upon chemical reduction is remarkably depressed by the smaller cation substitution through the enhancement of (Ti–O) bonds. From their x-ray absorption spectroscopic studies, they suggested that the incorporation of trivalent smaller cations into the bismuth sites near TiO_6 octahedra results in the enhancement of the lattice stability of the bismuth titanate. Such a stabilization of TiO_6 octahedra can be understood by a competition between adjacent (Bi–O) and (Ti–O) bonds as well as the contraction of the unit cell.

The effect of isovalent doping at the A-site was studied by substitution of Ce^{3+} at the bismuth site for various compositions of cerium in bismuth titanate host lattice $Bi_{4-x}Ce_xTi_3O_{12}$ ($x = 0.2$, 0.4, 0.6, and 0.8) (BCeT) [77]. The existence of two phases, (i) the Aurivillius bismuth titanate layer-structured phase and (ii) cubic pyrochlore phase ($Bi_2Ti_2O_7$), in Ce doped BIT ceramics was confirmed by x-ray diffraction studies. The pyrochlore phase was found to increase with an increase in cerium content in BIT, indicating that Ce is well substituted in the Bi site. It was noted that cerium ions can be incorporated into the bismuth titanate lattice in two oxidation states as they can exist in Ce^{3+} and Ce^{4+} oxidation states. In the 4+ oxidation state, it could substitute titanium at Ti^{4+} sites, whereas in oxidation state 3+ it could be incorporated at the Bi^{3+} site. However, as the ionic radius of Ce^{3+} (1.01 Å) is closer to that of Bi^{3+} (1.03 Å) than of the Ti^{4+} ion, the possibility of it occupying the Bi site is greater [77]. Morphological analysis carried out using SEM micrographs reveal that as the value of x in the composition $Bi_{4-x}Ce_xTi_3O_{12}$ increases, the grain size of these samples decreases. It was also noted that the grain size is more uniformly distributed in the sample with higher cerium content, which may be due to the fact that the fine grains that exist between large anisotropic grains become larger with higher x values. These observations are in agreement with those of Tang *et al* [78]. This suggests that the addition of a small amount of cerium controls the growth of grains in BIT. A change in the microstructure with doping concentration can be attributed to the inhibition of the Bi^{3+} volatilization by substitution of Ce^{3+} and is believed to be due to the formation of cubic pyrochlore phase bismuth titanate, as observed in XRD results. The variation of the dielectric constant as a function of temperature and frequency was studied in detail. It was observed that the ferroelectric phase transition temperature T_c shifts towards the lower temperature side with an increase in doping concentration. This may be a consequence of the incorporation of Ce^{3+} in the $Bi_4Ti_3O_{12}$ lattice, which causes crystal lattice distortion [79]. The value of the dielectric constant at the Curie temperature increases with the increase in doping level. The possible reasons for this could be the creation of structural defects due to the unique nature of cerium which changes its oxidation state from +3 to +4 and the presence of the secondary $Bi_2Ti_2O_7$ phase. As the doping concentration is increased, the ε_r versus temperature curves show a diffuse phase transition, which is the characteristic of a relaxor ferroelectric behaviour. The compounds show a temperature-dependent relaxation process. The activation energy for relaxation was found to be greater than that for conduction, indicating that the conductivity is not governed by the relaxation mechanism. The impedance plots suggest that the as-prepared samples follow Maxwell–Wagner (M-W) relaxation behaviour. The behaviour of conductivity with temperature suggests that the conduction mechanism is thermally activated at higher frequencies. The remanent polarization tends to increase with increasing doping content due to a decrease in the conductivity of the samples.

On similar grounds, isovalent doping at the B-site of bismuth titanate was studied by substituting Zr^{4+} at the Ti^{4+} site of BIT [80]. The Zr^{4+} ion is chemically more stable than the Ti^{4+} ion and has a larger ionic radius that can expand the pseudo-perovskite layers. The ionic state of titanium toggles between 3+ and 4+, thus causing conduction in the BIT ceramic. The modification of the B-site of bismuth titanate ceramics by Zr^{4+} ions reduces the conductivity due the electronic hopping between $Ti^{3+}$and Ti^{4+}. It also lowers the leakage current. Zirconium doped bismuth titanate ceramics were prepared successfully by solution-based combustion techniques and the addition of Zr^{4+} does not affect the lattice symmetry of BIT, but growth along the lattice plane (0010) is observed due to orthorhombic distortion with increasing doping level. The increase in the value of the dielectric constant and the decrease in the value of dielectric loss was observed with T_c remaining almost unchanged compared to undoped bismuth titanate. As the concentration of Zr^{4+} is increased, the conductivity due to the grain boundary dominates. The activation energy due to relaxation is found to be greater than that due to conduction, thus confirming that electrical conduction is not due to relaxation of the dipoles. The activation energy decreases with an increase in doping concentration due to orthorhombic distortion. Remanent polarization of the doped samples increases as the doping level of the zirconium is increased, which can be attributed to an increase in the orthorhombic distortion due to Zr^{4+} addition.

The effect of simultaneous doping in $Bi_4Ti_3O_{12}$ at the A- and B- sites, respectively, by the isovalant dopants (Ce^{3+} and Zr^{4+}) was studied [81]. The dopant concentration ($x = 0.6$) at both the A- and B- sites was chosen on the basis of the stable results obtained in the previous studies, when one dopant was used in the absence of the other with the same concentration. Also, the tolerance factor as calculated for simultaneous doping of bismuth titanate ceramics with Ce^{3+} and Zr^{4+} at the A- and B-sites, respectively, was found to be 0.86. This is in agreement with the tolerance factor deduced by the theory of crystal chemistry according to which an orthorhombic structure should have a tolerance factor lying between 0.71 and 0.91. Therefore, a new ceramic with composition $Bi_{3.4}Ce_{0.6}Ti_{2.4}Zr_{0.6}O_{12}$ (CeZrBIT) was synthesized using solution-based combustion technique. Its phase formation, morphology, and its effect on dielectric properties, ferroelectric properties, conduction mechanism, and relaxation behaviour as a function of temperature and frequency has been studied in detail. The x-ray diffraction pattern of the Ce–Zr co-doped BIT shows sharp peaks, which are indicative of the fact that the sample is completely crystalline. An extra peak is observed which corresponds to the pyrochlore bismuth titanate ($Bi_2Ti_2O_7$) phase indexed by JCPDS file no. 32-0118 and is attributed to the addition of Ce^{3+} (also reported in Ce doped BIT ceramics). In addition to this, the intensity of the peaks corresponding to (115) and (0010) in the XRD patterns of $Bi_{3.4}Ce_{0.6}Ti_{2.4}Zr_{0.6}O_{12}$ is found to increase compared to that of pure BIT ceramics. This can be ascribed to orthorhombic distortion in the BIT lattice due to Zr^{4+} addition (discussed for Zr doped BIT). Apart from this, no extra peaks are observed, indicating that Ce^{3+} and Zr^{4+} have been incorporated into the bismuth titanate lattice and no secondary phases are present. Figure 2.7(a) shows the variation of dielectric constant with temperature at 500 kHz for pure, Ce, Zr and Ce-Zr doped BIT ceramics. It is clearly evident that Ce substitution at Bi site decreases the Curie temperature whereas, for ceramics with Zr substitution at Ti site in BIT shows no effect on the transition temperature. The conductivity follows Jonsher's power law according to which the plateau region in the curve represents dc conductivity, and in the dispersion region at higher frequencies the conductivity is linearly dependent on the frequency. Jonsher's double power law can be represented as

$$\sigma(\omega) = \sigma_{dc} + A_1\omega^{s_1} + A_2\omega^{s_2}, \tag{2.10}$$

where A_1 and A_2 are pre-exponential factors, and s_1 and s_2 are exponents of frequency. Almond *et al* [82] suggested that such behaviour is indicative of the material behaving as a network of resistors (R) and capacitors (C). The network exhibits the dependence of ac conductivity on frequency in the various frequency domains as follows:

i. At low frequencies, where $R^{-1} > \omega C$, currents flow preferentially through the resistors and the conductivity is therefore frequency-independent.
ii. At high frequencies, where $R^{-1} < \omega C$, currents flow preferentially through the capacitors and the conductivity become frequency-dependent.
iii. At intermediate frequencies, where $R^{-1} \sim \omega C$, currents flow through both R and C.
iv. Power law dependences are observed in the frequency range where currents flow through both resistors and capacitors, i.e. when $R^{-1} \sim \omega C$.

It is observed that the *s* value decreases with an increase in temperature which indicates that in the *R–C* model the number of capacitive sites is decreasing [83]. This temperature dependence of the *s* value indicates that the conduction mechanism in these materials is not due to the intrinsic polarization process, but is a thermally activated phenomenon which is associated with the hopping of charged species across the charged defects. Nyquist plots for pure, Ce, Zr and Ce-Zr co-doped BIT ceramics at room temperature are shown in figure 2.7(b). The binding energy has been calculated using equation (2.7), which is defined as the energy required to remove an electron from one site to another. The decrease in slope with temperature is due to a decrease in binding energy. It is noted that binding energy decreases with an increase in temperature, which explains the increase in conductivity of the sample with temperature. In addition, it was observed that the values of binding energy for CeZrBIT are much higher than that of pure BIT for all temperatures, indicating that the conductivity decreases due to the simultaneous doping of Ce^{3+} and Zr^{4+} at the A- and B-sites of bismuth titanate. The values of binding energy can be used to determine the minimum hopping distance R_{min} by equation (2.8). It is characterized by a very low value (~10^{-10} m). It was observed that the minimum hopping distance increases with increasing temperature, which indicates the long range mobility of charge carriers. This can be attributed to the fact that under the action of induced electric field, there is a lack of restoring force governing the mobility of charge carriers. It can also be noted that the values of R_{min} for the doped sample are much lower than those of undoped bismuth titanate ceramics. Based on the CBH model, the ac conductivity data can be used to evaluate the density of states at the Fermi level $N(E_f)$ using equation (2.9). It is observed that $N(E_f)$ decreases with an increase in frequency for all temperatures and merges above 50 kHz. It is clear that $N(E_f)$ increases with temperature at lower frequencies and is independent of temperature at higher frequencies.

Improvement in the ferroelectric properties was observed due to doping the bismuth titanate ceramic with Ce^{3+} and Zr^{4+} simultaneously at the A- and B-sites [83]. A partially saturated loop with remanent polarization (P_r) = 1.25 μC cm^{-2} and coercive field (E_c) = 12 kV cm^{-1} is formed for doped ceramics. Improvement in the ferroelectric properties of doped BIT ceramic is believed to be due to a decrease in conductivity because of a reduction in the defect sites, which are created as a result of Bi^{3+} ion volatilization and oxygen vacancies due to instability in the oxidation state of Ti^{4+}.

Acceptor doping in $Bi_4Ti_3O_{12}$ may decrease the conductivity as it will compensate for the electrons produced due to oxidation of oxygen vacancies. Strontium doping in BIT at the bismuth site affects the dielectric properties and the conduction in the base compound. It was reported [84] that a small amount of Sr substitution in BIT reduces the ferroelectric to paraelectric transition temperature. The conductivity in Sr doped BIT showed two different effects on the grain and grain boundary conductivities. The conductivity of grains in the $Bi_4Ti_3O_{12}$ ceramic increased due to the addition of a small amount of Sr in the entire experimental temperature range, while the grain boundaries of 1 at% Sr doped $Bi_4Ti_3O_{12}$ exhibited lower conductivity than pure $Bi_4Ti_3O_{12}$ below ~380 °C and higher conductivity above ~380 °C. This diverse behaviour of conductivity in the two temperature regions is explained by the fact that the substitution of Sr for Bi increases the number of intra-grain oxygen vacancies resulting in higher conductivity.

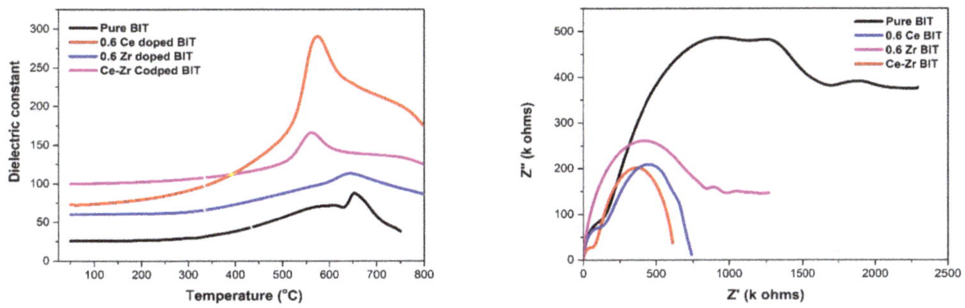

Figure 2.7. (a) Variation of dielectric constant with temperature for Ce, Zr, and Ce–Zr doped $Bi_4Ti_3O_{12}$ ceramics at a frequency of 500 kHz. (b) Room temperature Nyquist plots for Ce, Zr, and Ce–Zr doped $Bi_4Ti_3O_{12}$ ceramics.

At low temperatures, the Sr dopants play a hole-doping role in grain boundaries and consequently result in a higher resistance of Sr doped samples than that of pure BIT. At high temperatures, the Sr dopants induce more oxygen vacancies in grain boundaries to cause higher conductivity in Sr doped samples than in of pure BIT [84].

Structure refinements in $Bi_{4-x}Ba_xTi_{3-x}Nb_xO_{12}$ system prepared through solid-state reaction and structural features of relaxor Aurivillius phases were reported by Tellier *et al* [85]. They reported that in compounds with relaxor-like behaviour, the average structure is almost undistorted, close to the archetypal high temperature paraelectric phase, with a tolerance factor 0.996. Using transmission electron microscopy (TEM), they revealed some characteristic features of relaxors, such as micro-twinning and shearing-type defects, which arise from the existence of a compositional inhomogeneity and disorder at a local scale.

Magnesium doped bismuth titanate $Bi_4Ti_{3-x}Mg_xO_{12}$ ($x = 0.2$, 0.4, and 0.6) ceramics were prepared using a solution combustion technique [86]. The role of magnesium doping at the B-site of BIT was studied to understand its effect on the crystalline structure, ferroelectric properties, and electrical transport characteristics of Mg doped bismuth titanate as a function of temperature and frequency. It was observed that the addition of Mg^{2+} does not affect the lattice symmetry of the bismuth titanate and no secondary phases were observed. The value of the dielectric constant is found to increase and the value of dielectric loss to decrease with T_c shifting towards lower temperatures in Mg doped BIT ceramics compared to that of undoped bismuth titanate. As the concentration of Mg^{2+} is increased, the electrical conductivity due to the grain boundary dominates. The activation energy increases with an increase in doping concentration. The remanent polarization increases with an increase in Mg content. However, the *P–E* loops tend to become lossy due to Mg addition as a result of an increase in leakage current. This may be attributed to lattice distortion introduced in the unit cell.

Many researchers [87–89] reported the synthesis, and dielectric and ferroelectric properties of rare-earth doped BIT ceramics. They also reported that the doping of rare earths (La, Pr, etc) into BIT decreases the transition temperature. A diffuse phase transition (DPT) and relaxor-like behaviour were reported in the samples $Bi_{3.2}Nd_{0.8}Ti_3O_{12}$ [90] and $Bi_{3.2}Pr_{0.8}Ti_3O_{12}$ [91].

Rare-earth doped bismuth titanate $Bi_{4-x}R_xTi_3O_{12}$ (R = La, Nd, Sm, Gd) for $x = 0.75$ thin films deposited on $Pt/TiO_x/SiO_2/Si$ substrates by the chemical solution deposition (CSD) method using metallo-organic precursor solutions were studied by Yamada *et al* [92]. The crystallization behaviour, ferroelectric properties, and microstructures of these films were evaluated. Nd doped BIT thin film showed the best saturation properties of ferroelectrics with large $P_r = 21.6\ \mu C\ cm^{-2}$ and small $E_c = 80\ kV\ cm^{-1}$ from low applied voltages, among the rare-earth doped BIT thin films that were investigated in their studies [92]. The *P–E* loops for La, Nd, Sm, and Gd doped bismuth titanate thin films obtained by chemical solution deposition are shown in figure 2.8. Evidently, rare-earth substitution at the A-site in the perovskite layers enhances the ferroelectric properties of pure BIT radically. This improvement in the ferroelectric properties of BIT may be attributed to the rotation of TiO_6 octahedra in the *a–b* plane accompanied by a shift of the octahedron along the *a*-axis in the pseudo-perovskite layer [93]. This results in both the Bi_2O_2 layer as well as TiO_6 tetrahedron being considerably distorted.

B-site doping in bismuth titanate ceramics as well as in thin films has been studied for various isovalent and heterovalent dopants. Niobium doped bismuth titanate ($Bi_4Ti_{2.86}Nb_{0.14}O_{12}$) shows a piezoelectric coefficient $d_{33} = 20\ pC\ N^{-1}$ as compared to $d_{33} = 3.5\ pC\ N^{-1}$ for pure BIT [20]. Also, an increase in conductivity with the addition of acceptor dopants and a decrease in conductivity with the addition of donor dopants was observed. They also reported p-type conductivity in pure and acceptor doped BIT ceramics and suggested that the oxygen vacancies present in the sample due to sintering at a high temperature affects the conductivity. Shulman *et al* [94] observed three relaxation processes in the dielectric data in BIT, two of which combine at high frequencies (1 MHz) to produce an anomaly in the relative

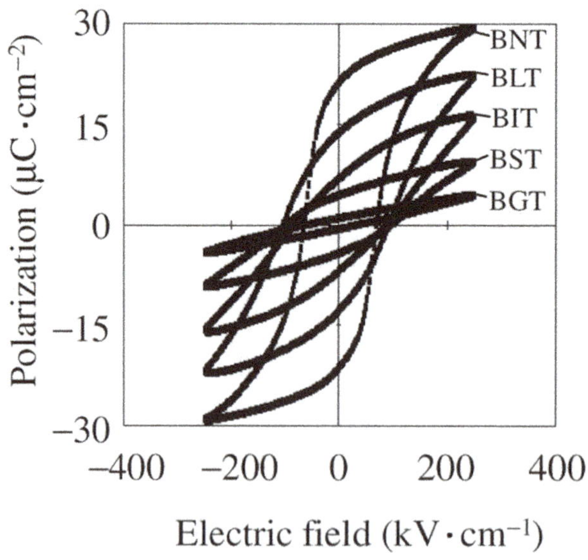

Figure 2.8. *P–E* hysteresis loops of rare-earth (La, Sm, Gd, Nd) doped BIT thin films prepared at 750 °C on $Pt/TiO_x/SiO_2/Si$ substrates by chemical solution deposition. Reproduced with permission from [92]. Copyright 2009 IOPscience.

permittivity that is centred at nearly 550 °C, which was observed by several researchers [34, 95–97]. This peak was attributed to phenomena such as structural changes, space charge relaxation, and oxygen vacancy movement [69, 70]. One set of peaks in the dielectric loss tangent was explained in terms of a Debye-type relaxation, which resulted from an ion-jump mechanism that occurred preferentially in the *a–b* plane. They did not observe dielectric loss peaks in Nb^{5+} doped BIT, as Nb^{5+} is believed to affect the relaxation through the suppression of oxygen vacancies.

Noguchi *et al* [98] reported that vanadium doping into BIT ceramics enables sintering at temperatures 100 °C–200 °C lower than that for the $SrBi_2Ta_2O_9$ system. They observed a large remanent polarization ($2P_r$) of over 40 $\mu C\ cm^{-2}$ and this value was shown to be much larger than that of the $SrBi_2Ta_2O_9$ system. They suggested that vanadium doped BIT could be an alternative material for nonvolatile random-access memories (NvRAMs) with large remanent polarization and lower processing temperatures. They also reported that the oxygen vacancies generated by inevitable bismuth vacancies would assemble at domain boundaries at relatively high temperatures, leading to a strong domain pinning in pure bismuth titanate. In contrast, vanadium doping could be effective for a decrease in oxygen vacancies and significantly reduce the influence of domain pinning, which would be responsible for the very large $2P_r$ found for vanadium doped BIT.

The electrical conduction behaviour of pure and vanadium doped BIT single crystals were investigated by Takahashi *et al* [99] using the complex impedance method. They reported that pure and vanadium doped BIT showed oxide ionic and p-type conduction. They suggested that bismuth and oxygen vacancies preferentially exist in the pseudo-perovskite blocks in BIT, and vanadium substitution effectively decreases the concentration of oxygen vacancies and holes in the pseudo-perovskite blocks.

Tungsten doped BIT and tungsten/manganese co-doped BIT prepared by a conventional solid-state reaction method were found to reduce oxygen vacancies, thus resulting in the reduction of ionic and electronic conductivities [100]. They show a p-type mixed conduction mechanism. Incorporation of W or Mn ions into the BIT lattice reduces the concentration of intrinsic oxygen vacancies and holes, resulting in a decrease of the electrical conductivity of W–Mn co-doped BIT. W doped and W/Mn co-doped BIT ceramics show similar activation energies and ionic-p-type mixed conduction mechanisms. The W doped BIT ceramics show higher resistivity and piezoelectric coefficients with decreased relative permittivity and dielectric loss compared to pure BIT. Meanwhile, the W/Mn co-doped ceramics exhibit larger pyroelectric and piezoelectric coefficients, and lower relative permittivity and dielectric loss compared to singly W doped ones. This increase can be attributed to an additional polarization change due to Mn/Ti–O defect dipoles superimposed on the spontaneous polarization change with temperature or electric field. The large decrease in dielectric loss is due to the domain wall motion being pinned by the defect dipoles. The effect of acceptor doping is evident in the build-up of an internal bias field, as evidenced by a horizontal offset in the polarization-field behaviour.

The dielectric and electrical properties of the pure and doped BIT $Bi_{4-x}La_xTi_3O_{12}$ ($x = 0.5$, 0.75, and 1) and $Bi_{4-y/3}Ti_{3-y}Nb_yO_{12}$ ($y = 0.015$, 0.03, and 0.06) ceramics prepared through the conventional solid-state reaction method were investigated by

Yao *et al* [101] from room temperature to 350 °C. They attributed the relaxation peaks in the plot of dielectric loss versus temperature for pure BIT to be associated with the motion of the oxygen vacancy. They found that the La^{3+} doping at the A-site favours improvement of the fatigue property and, in contrast, the Nb^{5+} doping at the B-site enhances the remanent polarization. This was attributed to the fact that appropriate La^{3+} doping at the A-site of the perovskite-type unit in BIT enhances the chemical stability of the oxygen vacancy by improving the height of the potential barrier for hopping and enhances the mobility of the domain by changing the domain structures. In contrast, the Nb^{5+} doping at the B-site induces the distortion of oxygen octahedral and reduces the oxygen vacancy concentration by a compensating effect so that it results in an enhancement of remanent polarization.

The ferroelectric and dielectric properties of pure BIT and Zr^{4+}, Hf^{4+}, and Nb^{5+} doped BIT at the B-site bulk ceramics prepared through the conventional solid-state reaction method were studied by Li *et al* [102]. They reported $2P_r$ values of 14.6 μC cm^{-2}, 23.6 μC cm^{-2}, 28.4 μC cm^{-2}, and 67.9 μC cm^{-2} for pure BIT, Zr^{4+}, Nb^{5+}, and Hf^{4+} doped BIT, respectively. They also reported that the enhancement of the P_r value has no certain relation with the doping ionic radius and the concentration of oxygen vacancy. The concentration and the activation energy of the oxygen vacancy are the two aspects that control the fatigue behaviour in B-site-substituted BIT ferroelectric materials. However, oxygen vacancies play an important role in fatigue endurance in doped BIT.

The microstructure, and dielectric and ferroelectric properties of $Bi_{4-y}Nd_yTi_{2.98}V_{0.02}O_{12.01}$ (with y = 0.16, 0.26, 0.36, and 0.56) ceramics prepared using the solid-state reaction method were reported by Tang *et al* [103]. The Curie temperature of compounds was found to decrease with increasing Nd concentration. In addition, these ceramic compounds exhibit saturated ferroelectric hysteresis loops with $2P_r$ = 26–31 μC cm^{-2} and E_c = 41–57.5 kV cm^{-1}. They observed the highest $2P_r$ of 31 μC cm^{-2} at E_c of 54 KV cm^{-1} for $Bi_{3.64}Nd_{0.36}Ti_{2.98}V_{0.02}O_{12.01}$.

Tripathy *et al* [104] reported the synthesis and characterization of $Bi_4Ti_{3-2x}M_xFe_xO_{12}$ (M = Ta, Nb) that retains the orthorhombic non-centrosymmetric structure of the parent BIT. They also reported the synthesis of two new compounds of this family, $Bi_2Sr_2Nb_2RuO_{12}$ and $Bi_2SrNaNb_2RuO_{12}$, which are analogous to $B_{i2}Sr_2Nb_2TiO_{12}$ possessing a tetragonal (*I4/mmm*) Aurivillius structure.

2.3 Five-layered oxides (m = 5)

2.3.1 Structure

Five-layered bismuth titanate ceramics were first synthesized by Subbarao and Aurivillius in 1962 [105, 106]. In five-layered Aurivillius oxides such as $Ba_2Bi_4Ti_5O_{18}$, $Pb_2Bi_4Ti_5O_{18}$ and $Sr_2Bi_4Ti_5O_{18}$, the perovskite block $((Ba/Pb/Sr)_2Bi_2Ti_5O_{16})^{2-}$ is interleaved between the $(Bi_2O_2)^{2+}$ layer as shown in figure 2.9. Here the A-site is occupied by Ba/Pr/Sr along with Bi and Ti^{4+} ions residing at the B-site of the pseudo-perovskite block. Subbarao also suggested that Bi ions and Ti ions are suitable for the A- and B-sites in the perovskite block and increases the size restriction with an increase in the value of m. Compounds with perovskite blocks with an even number of BO_6 octahedra show no spontaneous polarization along the c-axis because of the existing mirror symmetry. However, a small degree of spontaneous polarization can be observed in the compounds with perovskite blocks with an odd number of BO_6 octahedra along the *c*-axis [28, 107, 108]. Some researchers also observed that the ferroelectric and structure stability is closely related to the value of *m*. A saturated hysteresis loop with a small coercive field is obtained as the value of *m* increases because in certain component systems it leads to an increase in the lattice parameter in the *a*–*b* plane and also an increase in the strain arising due to the size mismatch between the $(Bi_2O_2)^{2+}$ and perovskite layers, making the structure unstable [109]. The positional disorder of cations on the A- or B-site of the perovskite blocks is responsible for the of relaxor-like behaviour. The low values of the dielectric constant at the transition temperature and the small shift in the maximum of the transition temperature to the lower temperature side indicates a weak interaction between the dipole moments as observed in another Ba and Pb based layered perovskite [110]. It is believed that Ba^{2+}/Pb^{2+} ions substitute not only the A-site of the perovskite blocks, but also enter into the $(Bi_2O_2)^{2+}$ layers that create a local charge imbalance. The reason for dielectric dispersion is the formation of defects due to an inhomogeneous distribution of Ba/Pb, which results in thermally activated conduction of mobile ions and/or other defects [111]. The peak position is strongly dependent on the frequency and indicates the possibility of relaxor behaviour. Therefore, $Ba_2Bi_4Ti_5O_{18}$ and $Pb_2Bi_4Ti_5O_{18}$ compounds in the five-layered family are relaxor ferroelectric, whereas their $Sr_2Bi_4Ti_5O_{18}$ counterpart behaves like a normal ferroelectric with a sharp transition at the Curie temperature (figure 2.9).

$Sr_2Bi_4Ti_5O_{18}$ (SBT18) exhibits an orthorhombic (pseudo-tetragonal) crystal structure at room temperature with $a = b = 5.461$ Å, $c = 48.8$ Å. The variation in dielectric constant with temperature measured on these ceramics showed a ferro-electric to paraelectric phase transition occurring at 285 °C. Above the transition temperature, SBT18 shows a tetragonal structure and the dielectric constant follows the Curie–Weiss law [105]. Fang *et al* described the formula of $Ba_2Bi_4Ti_5O_{18}$ in the form of $Bi_4Ti_3O_{12}{\cdot}2BaTO_3$ and suggested that this compound has a layered structure and shows ferroelectric properties [113]. The orthorhombic (pseudo-tetragonal) space group *B2eb* structure of the five-layer Aurivillius phases $A_2Bi_4Ti_5O_{18}$ (A = Ba, Pb, Sr, Ca) was refined from powder diffraction data using the Rietveld refinement and it was observed that the structure consisted of a $[Bi_2O_2]^{2+}$ layer interleaved with a perovskite-like $[A_2Bi_2Ti_5O_{16}]^{2-}$ block [26]. In 1962 Aurivillius synthesized $Ba_2Bi_4Ti_5O_{18}$ (BBT18) and reported that BBT18 has five-layer perov-skite structures with $T_c = 329$ °C, at which temperature the compound exhibits a pseudo-tetragonal to tetragonal crystal structure and the ferroelectricity and electrical properties of $Ba_2Bi_4Ti_5O_{18}$ single crystals were also reported. There were conflicting reports about the existence of a single phase sample of $Ba_2Bi_4Ti_5O_{18}$, Subbarao reported that this is a mixture of the four-layer oxide $BaBi_4Ti_4O_{15}$ and $BaTiO_3$, while Aurivillius claimed that $Ba_2Bi_4Ti_5O_{18}$ could be obtained as a single phase [105]. Ikegami *et al* reported the preparation of a mixture of $SrTiO_3$ and BIT possessing perovskite slabs with increasing Sr content and exhibiting good dielectric and ferroelectric properties [114]. Subbarao reported the ferroelectric to paraelectric phase transition of BBT18 and SBT18 at 310 °C and 285 °C, respectively [115]. In 1994 Zou *et al* investigated the temperature behaviour of the vibrational mode in

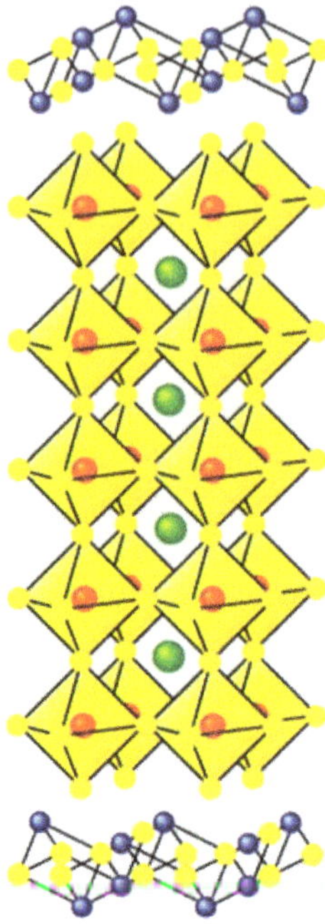

Figure 2.9. The structure of five-layered Aurivillius oxides. Reproduced from [112].

five-layered BLSF compound $Sr_2Bi_4Ti_5O_{18}$ with the help of Raman spectra. The soft mode frequency of Raman spectra at a higher temperature decreases with an increase in temperature and disappears above the Curie temperature. The soft mode frequency is dependent on the displacement of the Bi^{3+} ion and ferro- to paraelectric phase transition [116]

2.3.2 The synthesis of five-layered ceramics and thin films and their properties

The preparation of five-layered Aurivillius oxides using a conventional solid-state reaction method involves the heating of precursor materials at higher temperatures. This results in inhomogeneous particle size, bigger crystallite size, phase impurities, etc. These problems can be overcome using low temperature wet chemical methods. Synthesis through the sol–gel method has exceptional control over its composition and homogeneity at the molecular level, but the resulting precipitates are amorphous in nature. Moreover, high purity yield with fine crystallite size can be obtained by the co-precipitation technique. Nonetheless, since this synthesis technique is carried out in an aqueous environment, it induces water molecules into the samples, thus deteriorating their electrical properties. The solution combustion technique results in fine powders with particle sizes ranging from nanometres to a few micrometres. However, the powders obtained through combustion synthesis are porous due to the high flame temperature.

Five-layered Aurivillius oxide $Ba_2Bi_4Ti_5O_{18}$, $Pb_2Bi_4Ti_5O_{18}$, and $Sr_2Bi_4Ti_5O_{18}$ ceramics were synthesized using the sol–gel technique [117]. For all the compounds, the calcination time and temperature were optimized by studying the phase formation at different temperatures and times of calcination. It was observed that these five-layered compounds, when calcined at 700 °C for 2 h, were amorphous in nature. As the calcination time was increased to 4 h the samples did attain crystallinity but the presence of secondary phases was observed. Pure phase compounds of $Ba_2Bi_4Ti_5O_{18}$, $Pb_2Bi_4Ti_5O_{18}$, and $Sr_2Bi_4Ti_5O_{18}$ were obtained at a calcination temperature of 750 °C for 4 h. The sintering temperature was optimized by studying the dielectric, electrical, and ferroelectric properties for the samples sintered at three different temperatures 850 °C, 950 °C, and 1050 °C. It was observed that the samples sintered at 1050 °C possessed a higher density with enhanced dielectric and ferroelectric properties. Figure 2.10 shows the variation of the dielectric constant with temperature for BBT18, PBT18, and SBT18 ceramics sintered at different temperatures. It is observed that the dielectric constant increases with an increase in sintering temperature. Table 2.2 shows the crystallite size, grain size, density, dielectric constant, and dielectric loss at room temperature (RT), Curie temperature (T_c), activation energy (E_a), conductivity (σ), remanent polarization (P_r),

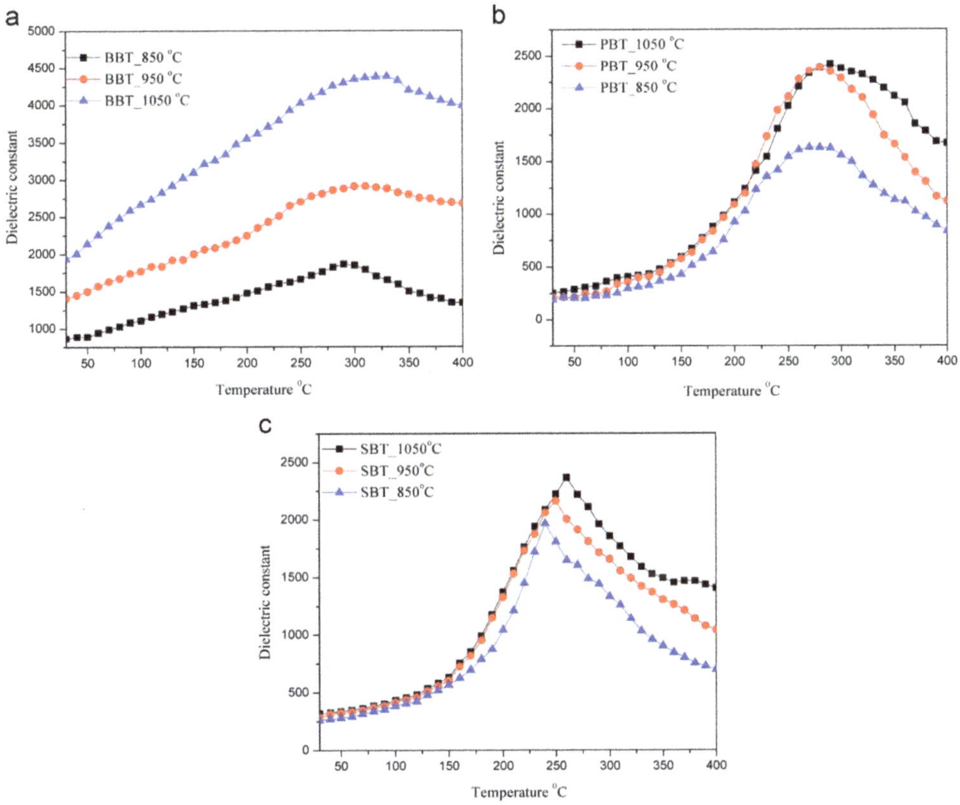

Figure 2.10. Variation in the dielectric constant as a function of temperature at 1 kHz for (a) $Ba_2Bi_4Ti_5O_{18}$, (b) $Pb_2Bi_4Ti_5O_{18}$, and (c) $Sr_2Bi_4Ti_5O_{18}$ ceramics sintered at different temperatures (850 °C, 950 °C, and 1050 °C) prepared using the sol–gel method. Reproduced from [117], Copyright (2017), with permission from Elsevier.

Table 2.2. Crystallite size, grain size, density, dielectric constant, and dielectric loss at RT, Curie temperature (T_c), activation energy (E_a), conductivity (σ), remanent polarization (P_r), and coercive field (E_c) for $Ba_2Bi_4Ti_5O_{18}$, $Pb_2Bi_4Ti_5O_{18}$, and $Sr_2Bi_4Ti_5O_{18}$ ceramics. Reproduced from [117], Copyright (2018), with permission from Elsevier.

Composition with sintering temperature (°C)	Crystallite size (nm)	Grain size (nm)	Density %	ε_r at RT	T_c °C	tanδ at RT	E_a (eV)	σ (Sm^{-1}) RT 100 kHz	$2P_r$ (μC cm^{-2})	$2E_c$ kV cm^{-1}
$Ba_2Bi_4Ti_5O_{18}$	36									
850		168	80	867	291	0.030	0.18	1.4E-9	0.04	3.4
950		222	90	1409	302	0.039	0.23	5.5E-11	0.07	3.6
1050		442	93	1937	326	0.049	0.27	2.3E-12	0.16	5
Co-precipitation				300 [119]	350 [119]	0.26	0.40	7.2E-11	0.054 [119]	
Solid-state reaction				300 [120]	324 [120]				0.7 [121]	
$Pb_2Bi_4Ti_5O_{18}$	31									
850		230	75	195	269	0.10	0.21	3.1E-8	0.14	5.2
950		360	91	216	278	0.16	0.25	4.8E-9	0.22	5.3
1050		398	95	252	292				0.42	12.8
Co-precipitation		398	95	220 [119]	309 [119]				0.25 [119]	
Solid-state reaction				100 [107]	330 [107]				0.8 [122]	
$Sr_2Bi_4Ti_5O_{18}$	54									
850		139	82	262	239	0.038	0.16	1.2E-7	0.07	4.2
950		297	91	288	250	0.042	0.19	7.2E-7	0.10	7.8
1050		540	94	319	259	0.237	0.39	5.0E-8	0.22	16.2
Co-precipitation				102 [119]	260 [119]				0.062 [119]	
Solid-state reaction				70, 120 [107]	285 [107]					

and coercive field (E_c) for $Ba_2Bi_4Ti_5O_{18}$, $Pb_2Bi_4Ti_5O_{18}$, and $Sr_2Bi_4Ti_5O_{18}$ ceramics sintered at different temperatures (850 °C, 950 °C, and 1050 °C) prepared using the sol–gel route and those reported for the solid-state reaction and co-precipitation synthesis techniques.

The effect of the synthesis route on the phase formation, microstructure, and dielectric, electrical, and ferroelectric properties of five-layered bismuth titanate ceramics obtained by different wet chemical routes was investigated by Dubey *et al* [119]. $Ba_2Bi_4Ti_5O_{18}$ (BBT18), $Pb_2Bi_4Ti_5O_{18}$ (PBT18), and $Sr_2Bi_4Ti_5O_{18}$ (SBT18) compounds were prepared by three different routes: sol–gel, co-precipitation, and solution combustion synthesis with optimization of calcination and sintering temperatures. The powders obtained by optimizing the calcination temperature were sintered at 1050 °C for 4 h. An attempt was made to understand the role of the adopted synthesis route and its processing parameters in tailoring various electrical and ferroelectric properties of these five-layered compounds for applications in memory devices.

A comparative study of the results obtained from $A_2Bi_4Ti_5O_{18}$ (A = Ba, Pb, Sr) ceramics synthesized by different processing routes, such as solution combustion, co-precipitation, and sol–gel, were reported by Dubey *et al* [119]. Their study suggests that sol–gel-synthesized $A_2Bi_4Ti_5O_{18}$ (A = Ba, Pb, Sr) ceramics possess a higher dielectric constant and remanent polarization, and low dielectric loss due to the lower conductivity in these samples as a result of their higher density. The morphology of the ceramics synthesized by three different routes was studied using SEM micrographs. It was observed that all the as-prepared five-layered Aurivillius oxides possess plate-like grain morphology, which is the characteristic of the layered structure. The grain size of the compounds prepared by the sol–gel synthesis technique is smaller and the grain size distribution is more uniform compared to the compounds prepared by the other two routes. The reason for the smaller grain size of the compounds obtained by the sol–gel route is due to its lower processing temperature resulting in smaller crystallite size. A uniform grain size distribution is characteristic of the sol–gel process. Figure 2.11 shows SEM images of $A_2Bi_4Ti_5O_{18}$ (A = Ba, Pb, Sr) ceramics obtained by three different wet chemical routes. Moreover, it is also observed that the ceramics obtained through the solution combustion synthesis route are porous as compared to the other two chemical routes (co-precipitation and sol–gel). Porosity is introduced in these samples during the combustion process due to the high flame temperature. Even though the solution combustion technique is cost effective and less time consuming, it produces porous powders after the combustion process. As the flame temperature during the combustion process is very high, it creates defect sites, which increases the dielectric loss. Hence, the lower density and defect sites in these ceramics are believed to be the main reasons for the higher conductivity, thus degrading its dielectric and ferroelectric properties. The co-precipitation technique requires multiple washings with ethanol and distilled water, due to which there is a probability of moisture retention in the ceramic, which might degrade the electrical properties. Barium bismuth titanate synthesized using sol–gel route is found to have a high dielectric constant, low dielectric loss, and lower conductivity compared to the other members of the

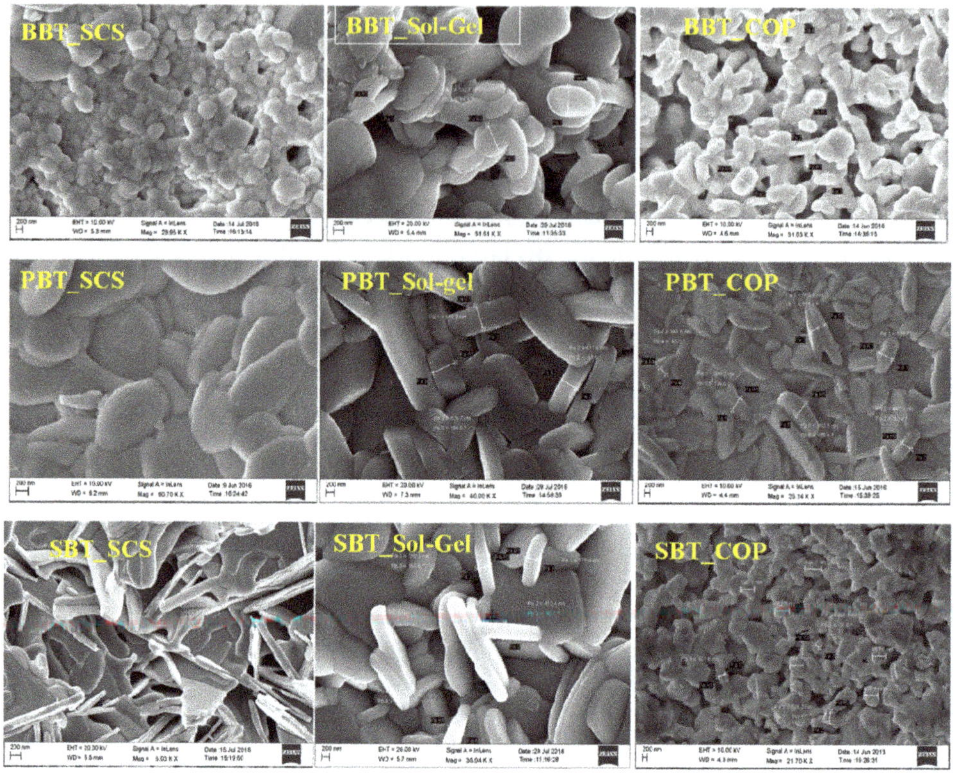

Figure 2.11. Surface morphology of (a) BBT18, (b) PBT18, and (c) SBT18 ceramics prepared using the solution combustion, co-precipitation, and sol–gel methods at a sintering temperature of 1050 °C. [118] (2018) Copyright. With permission of Springer.

five-layered Aurivillius compounds. Lead bismuth titanate possesses higher remanent polarization compared to the other members of the group due to the higher polarizability of Pb^{2+} ions. Hence, it can be concluded that five-layered Aurivillius oxides $A_2Bi_4Ti_5O_{18}$ (A = Ba, Pb, Sr) synthesized by the sol–gel route exhibit better dielectric, ferroelectric, and electrical properties.

The dielectric behaviour of the five-layered perovskite oxides $Sr_2Bi_4Ti_5O_{18}$ and $Pb_2Bi_4Ti_5O_{18}$ were studied by Onodera *et al* [109]. PBT18 exhibits a diffuse and broad anomaly around the transition temperature whereas SBT18 clearly shows a shoulder-like phase transition. The transition temperatures show a linear dependence on the ionic radii of the A-site ions in these layered perovskites. The transition temperature increases with an increase in the ionic radius of the A-site cation. Ferrer *et al* [122] used a mechano-chemical activation method for the synthesis of the Aurivillius phases $Bi_4SrTi_4O_{15}$ and $Bi_4Sr_2Ti_5O_{18}$. Compared to the solid-state reaction route, in the mechano-chemical activation method a significant decrease in the temperature of synthesis and the reaction time were observed by supplying mechanical energy to the initial reagents. Further, thermal annealing increases the crystallite and particle size. The products were characterized by Raman spectroscopy, x-ray diffraction, thermal analysis, and scanning electron microscopy.

Miyayama *et al* [123] reported the electrical anisotropy of $m = 2, 4, 5$ belonging to the bismuth layer-structured ferroelectric compounds ($PbBi_2Nb_2O_9$, $PbBi_4Ti_4O_{15}$, and $Pb_2Bi_4Ti_5O_{18}$) and discussed their relationship with conductivity and the number of octahedra in perovskite block m. The compounds with an odd number of m show larger anisotropy compared to the compounds with an even number of m. DC conductivity increases with an increase in the number of layers. Spontaneous polarization along the c-axis was not observed for compounds with an even number of m but it was observed for compounds with an odd m number [46]. Zhang *et al* prepared $Sr_2Bi_4Ti_5O_{18}$ thin film on a (100)-STO single crystal substrate using PLD. All the films had single phases confirmed by HRTEM and electron diffraction. The bismuth-layered perovskites showed octahedral tilting and had a high dielectric constant. The electrical and ferroelectric properties of $Pb_2Bi_4Ti_5O_{18}$ were discussed by Yi *et al* in 1998 [124]. Irie *et al* [125] synthesized $Ba_2Bi_4Ti_5O_{18}$ with a high dielectric constant and 360 °C transition temperature. They reported that due to the layered structure, $Ba_2Bi_4Ti_5O_{18}$ compounds showed layered anisotropy in conductivity, spontaneous polarization, and dielectric permittivity.

Nanocrystalline powder of $Sr_2Bi_4Ti_5O_{18}$ was synthesized using a citrate assisted sol −gel synthesis technique. SBT18 ceramics with grain sizes ranging from 93 nm to 1.43 μm were prepared by varying the sintering time and temperatures (800 °C−1150 °C). The grain-dependent dielectric, ferroelectric, and piezoelectric properties of these ceramics were studied. Ceramics with smaller grains showed a diffuse phase transition and higher Curie temperatures. T_c was found to decrease with the increase in grain size. This dependence of T_c on grain size was in accordance with extended Landau phenomenological theory. The SBT18 ceramics sintered at 1100 °C with an average grain size of 645 nm were found to show superior ferroelectric, dielectric, and piezoelectric properties with $2P_r = 16.4\ \mu C\ cm^{-2}$, coercive field $E_c = 38\ kV\ cm^{-1}$, piezoelectric coefficient $d_{33} = 22\ pC\ N^{-1}$, and planar electromechanical coupling coefficient $k_p = 14.8\%$ [126]. The variation of d_{33} and k_p with grain size is shown in figure 2.12.

High-power piezoelectric characteristics were studied in the layer-structured ferroelectric $MnCO_3$-doped $(Sr_{0.7}Ca_{0.3})_2Bi_4Ti_5O_{18}$ (SCBT0.3 + Mn x wt. % where $x = 0$, 0.2 and 0.4) ceramics. The poling treatment of SCBT0.3 requires a relatively high temperature of about 200 °C because of the large coercive field above 85 kV cm^{-1} at room temperature. At small-amplitude vibration, the electromechanical coupling factor k_{33} was found to be 0.136 for pure SCBT0.3 and 0.164 in SCBT0.3 + Mn 0.2 wt% ceramics poled at 200 °C. The doped Mn ions are very effective in improving the poling treatment and promoting the domain switching of SCBT0.3, increasing both k_{33} and Q_m [127].

Highly textured $Ba_2Bi_4Ti_5O_{18}$ ceramics were prepared by spark plasma sintering (SPS). The room temperature x-ray diffraction plots of the ceramics revealed that the majority of the XRD peaks correspond to the ferroelectric phase space group *B2cb*, while a few peaks correspond to the paraelectric phase space group *I4/mmm*. A diffuse phase transition around 240 °C was observed in the sample measured along the direction perpendicular and parallel to the direction of pressing. The slim ferroelectric *P–E* hysteresis loops, the presence of extra switching current peaks in

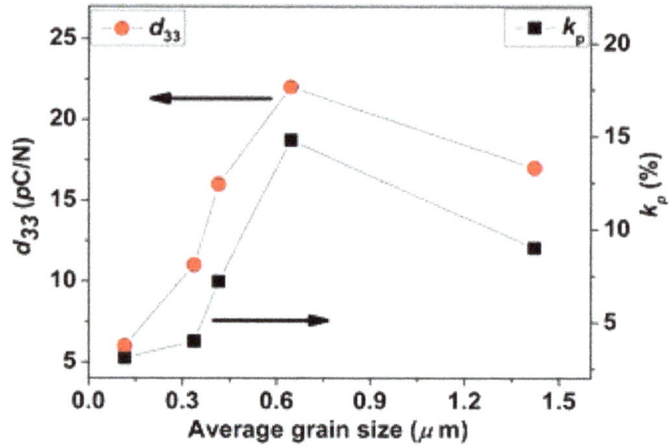

Figure 2.12. Grain size-dependent piezoelectric coefficient (d_{33}) and electromechanical planar coupling coefficient (k_p) for SBT18 ceramics at room temperature (27 °C). [126] (2016) Copyright. With permission of Springer.

the *I–E* loop, and the non-zero d_{33} piezoelectric coefficient (~ 0.4 ± 0.1 pC N^{-1} for the BBT18 in the direction perpendicular to pressing direction of the ceramic) can be considered as a confirmation of the relaxor ferroelectric state in the samples. The recoverable energy density (0.41 ± 0.01 J cm^{-3}) of $Ba_2Bi_4Ti_5O_{18}$ ceramics in the perpendicular direction to the SPS pressing are found to be comparable to that of Pb $(Mg_{1/3}Nb_{2/3})O_3$ based ceramics. These results suggest that SPS BBT18 ceramics might be promising materials for energy storage applications [128].

Ferroelectric thin films of $Sr_2Bi_4Ti_5O_{18}$ were grown on Pt/TiO_2/SiO_2/Si substrate using pulsed laser deposition (PLD) with and without a buffer layer and bottom electrode of $SrRuO_3$. $SrRuO_3$ was chosen as a material for the buffer to grow epitaxial $Sr_2Bi_4Ti_5O_{18}$ films due to its similar crystal structure and small mismatch with the lattice constants of $Sr_2Bi_4Ti_5O_{18.}$ XRD measurements reveal that *c*-axis oriented $Sr_2Bi_4Ti_5O_{18}$ films were grown on $SrRuO_3$ seeded Pt/TiO_2/SiO_2/Si substrates, whereas polycrystalline films were grown without a buffer layer on Pt/TiO_2/SiO_2/Si substrates. The polycrystalline films exhibit better ferroelectric properties with a higher value of remanent polarization and excellent retention characteristics after 10^5 sec [129]. SBT18 films were also deposited on the Pt/TiO_2/SiO_2/Si substrate using the sol–gel method. These films were random-oriented with a morphology of rod-like grains. The remanent polarization ($2P_r = 24$ μC cm^{-2}) and coercive field ($E_c = 68$ kV cm^{-1}) of the SBT18 films was found to be much higher than that reported for films deposited using pulsed laser deposition. The films with Sr-deficient and Bi-excess composition showed a very large remanent polarization of 35.5 μC cm^{-2}. Moreover, both these films showed high fatigue resistance against continuous switching up to 3×10^{10} cycles and excellent charge-retention ability up to 3×10^4 s [130].

$Ba_2Bi_4Ti_5O_{18}$ thin film was deposited on a Pt/Ti/SiO_2/Si substrate using chemical solution deposition. A high value of the room temperature dielectric constant of 260 and a low leakage current of 9.4 × 106 A cm^{-2} were obtained. A slim *P–E* loop was obtained with a large maximum spontaneous polarization and small remnant polarization. The obtained slim electric field polarization curve shows a large energy storage density of 22.6 J cm^{-3} and a high storage efficiency of 71.3% in the $Ba_2Bi_4Ti_5O_{18}$ film capacitors. The high cyclic fatigue endurance of the $Ba_2Bi_4Ti_5O_{18}$ thin films makes them a potential candidate for applications in dielectric energy storage capacitors [131].

The dielectric electrical and ferroelectric properties of barium bismuth titanate $Ba_2Bi_4Ti_5O_{18}$ single crystals grown via the melting–slow-cooling method were studied separately along the *a*- (or *b*-)axis and the *c*-axis. The dielectric permittivity was found to be 2000 at a frequency of 1 MHz measured along the *a*- (or *b*-)axis at a Curie temperature (360 °C), which was eight times higher than that along the *c*-axis. The dc conductivity along the *a*- (or *b*-)axis was roughly one order of magnitude greater than that along the *c*-axis in the temperature range of 350 °C–700 °C. The BBT18 single crystal had a saturated remanent polarization (P_r) of 12 μC cm^{-2} and a saturated coercive field (E_c) of 30 kV cm^{-1} along the *a*- (or *b*-)axis, and corresponding values of P_r = 0. 85 μC cm^{-2} and E_c = 8.1 kV cm^{-1} along the *c*-axis. The BBT18 single crystal shows large anisotropies in dielectric permittivity, conductivity, and spontaneous polarization because of the layered structure [132].

2.3.3 The effect of ion substitution in five-layered Aurivillius oxides

Barium doping in $Sr_2Bi_4Ti_5O_{18}$ tends to lower the Curie temperature to a minimum value of 150 °C up to a doping concentration of x = 0.75, which then increases on further increasing the doping content. However, the remanent polarization and coercive field decreased due to barium substitution. Also, the relaxor behaviour becomes more pronounced for Ba < 0.75, but beyond that relaxor characteristics weaken [133].

$Sr_{2-x}(Li, Ce)_{x/2}Bi_4Ti_5O_{18}$ ceramics synthesized by conventional ceramic processing were studied by Chen *et al* [134]. Their composition-dependent structure and electrical properties were investigated in the study. Pure phase Aurivillius oxides of $Sr_{2-x}(Li, Ce)_{x/2}Bi_4Ti_5O_{18}$ ceramics with uniform grain size were obtained. La, Ce substitution in $Sr_2Bi_4Ti_5O_{18}$ ceramics tends to monotonously increase the Curie temperature from 286 °C for pure SBT18 to 299 °C for $Sr_{2-x}(Li, Ce)_{x/2}Bi_4Ti_5O_{18}$ at $x/2$ = 0.05. The temperature dependence of the dielectric constant for different concentrations of La and Ce and the variation of T_c with doping concentration (inset) is shown in figure 2.13. The increase in Curie temperature due to (La, Ce) substitution may be attributed to a decrease in the tolerance factor. An increase in the values of remanent polarization is also observed in La, Ce substituted SBT18. The higher valency of La and Ce than that of Sr may be the cause of the higher ferroelectric distortion [134]. It was observed that the simultaneous doping of (Li, Ce) in SBT18 ceramic increases its Curie temperature and the remanent polarization. The SBT18 ceramics with an La, Ce composition of $x/2$ = 0.0125 possesses a maximum value of T_c = 299 °C, P_r = 9.3 μC cm^{-2}, and d_{33} = 26 pC N^{-1}. The improvement in ferroelectric properties may be attributed to the fact that the donor dopant creates cation vacancies, facilitating easier domain wall movement.

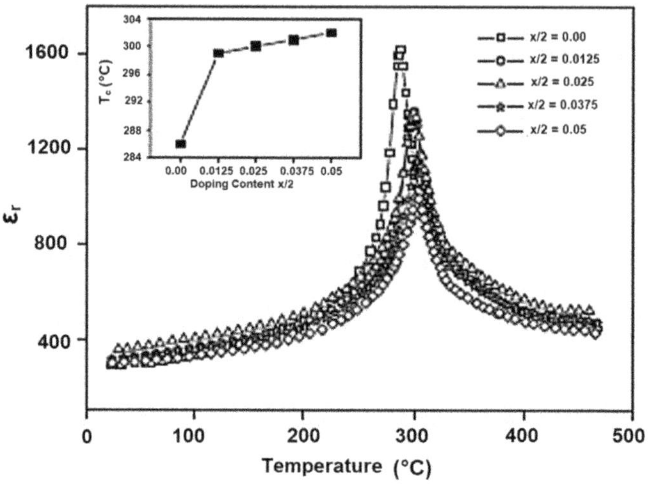

Figure 2.13. The temperature dependence of a dielectric constant for different concentrations of La and Ce doped SBT18. The inset shows variation of T_c with doping concentration. Reproduced from [134] with permission from World Scientific.

The effect of gadolinium (Gd) substitution in $Sr_2Bi_4Ti_5O_{18}$ ceramics synthesized using a conventional solid-state reaction on the dielectric, ferroelectric, and piezoelectric properties of the ceramics was studied by Chen *et al* [135]. It was observed that the addition of Gd^{3+} dopant decreases the Curie temperature (T_c) and lowers the dielectric loss (tanδ) of SBT18 ceramics. Also, Gd doped SBT18 ceramics with appropriate Gd content enhanced the ferroelectric and piezoelectric properties of the SBT18i ceramics. The piezoelectric constant (d_{33}) for the composition $Sr_2Bi_{3.9}Gd_{0.1}Ti_5O_{18}$ ceramic was found to be at a maximum with 22 pC N^{-1}.

Dysporium (Dy) doped $Sr_2Bi_4Ti_5O_{18}$ ceramics for different composition $Sr_2Bi_{4-x}Dy_xTi_5O_{18}$ (SBDT-x, x = 0.00, 0.003, 0.01, 0.05, 0.15, and 0.20) were prepared by a conventional solid-state reaction method [136]. Dy substitution at the Bi site in SBT18 ceramics is found to lower the Curie temperature and dielectric loss with an increase in Dy content. However, the dielectric constant remanent polarization increases initially with an increase in Dy concentration and reaches a maximum value of $2P_r$ = 20 μC cm^{-2} at E_c = 44.2 kV cm^{-1} for x = 0.01. It decreases on further increasing the Dy content. This may be attributed to a change in space charge content and lattice distortion induced due to Dy^{3+} substitution.

Nd-modified SBT1818 (SBNT) thin films with composition ($Sr_2Bi_{3.9}Nd_{0.1}Ti_5O_{18}$) were deposited on $Pt/Ti/SiO_2/Si$ (100) substrates using a sol–gel method [137]. The morphology of these thin films was studied using AFM micrographs. It was observed that these sol–gel derived films were randomly oriented and possessed rod-like grains. The ferroelectric properties of Nd doped SBT18 thin films were evaluated using P–E loops. The remanent polarization ($2P_r$) and coercive field (E_c) of the SBNT films were found to be 30 μC cm^{-2} and 55 kV cm^{-1}, respectively. This value of $2P_r$ was much higher than the reported value for SBT18 films obtained by pulsed laser deposition. Interestingly, the as-deposited SBNT films also showed higher fatigue resistance against continuous switching up to 3×10^9 cycles and an exceptional charge-retention capacity of 3×10^4 s.

Calcium doped SBT18 thin films $Ca_xSr_{2-x}Bi_4Ti_5O_{18}$ (CSBT18, x = 0–15 mol%) were fabricated using the sol–gel technique [138]. An optimum amount of Ca doping in SBT18 films was found to improve its ferroelectric performance. Morphological studies carried out using SEM micrographs revealed that the Ca doped SBT18 films with a doping content of 10 mol% were dense and crystalline with a uniform grain size distribution. XPS analysis of Ca doped SBT18 film shows that films with 10 mole% of Ca possess the lowest Ti^{3+} content, lower oxygen vacancies but higher remanent polarization $2P_r = 33.5\ \mu C\ cm^{-2}$, and almost double the value of the coercive field reported for other dopants, $2E_c = 182.8\ kV\ cm^{-1}$. In addition, the leakage current density was found to be $1.67 \times 10^{-7}\ A\ cm^{-2}$. The Ca doped SBT18 film with 10 mol% Ca also showed improved dielectric properties, with a dielectric constant $\varepsilon_r = 795$ and lower dielectric loss ($\tan\delta = 0.03$) at a frequency of 10^5 Hz. The dielectric tunability was found to be $T = 18.6\%$ at 200 kV cm^{-1} and 500 kHz.

To summarize, odd number layered (m = 1, 3, 5) Aurivillius oxide ceramics and thin films were studied by various researchers. They were found to show polarization along the a–b axis as well as along the c-axis. However, the films that are a-axis oriented possess much larger polarizations compared to those oriented along the c-axis. In addition, the Curie temperature of these compounds is found to decrease with an increase in the number of layers. Single-layered Aurivillius oxides BWO and BMO prepared using the solid-state reaction method are very high T_c (~900 °C) ferroelectric materials and they possess multiple phase transitions, such as ferroelectric to ferroelectric, ferroelectric to paraelectric, and reconstructive high temperature transition. Three-layered bismuth titanate $Bi_4Ti_3O_{12}$ ceramics have been prepared by conventional solid-state reaction and various other wet chemical routes and mechano-activation techniques. Their Curie temperature is reported to be ~ 675 °C. The low temperature synthesis of three-layered BIT using the sol–gel technique is reported. However, BIT ceramics synthesized using the conventional solid-state reaction route show better ferroelectric, piezoelectric, and pyroelectric properties than those synthesized by other routes. The dielectric, ferroelectric, piezoelectric, and pyroelectric properties of bismuth titanate thin films obtained by the pulsed laser deposition and chemical vapour deposition methods were also studied. Doping in bismuth titanate at the A-site and B-site by various types of dopants, such as isovalent and heterovalent dopants, was investigated by various researchers. Donor doped BIT using the pentavalent dopant Nb^{5+} at the B-site was also found to show remarkable improvement in the ferroelectric properties of bismuth titanate thin films with $2P_r$ ~ 67 $\mu C\ cm^{-2}$. Simultaneous doping at the A- and B-sites of bismuth titanate did reduce the electrical conductivity in ceramics as well as thin films, thus improving their ferroelectric and piezoelectric responses. The highest value of the pyroelectric coefficient ~57 $\mu C\ m^{-2}$ K was reported in W–Mn co-doped BIT ceramics. The five-layered Aurivillius oxides $Ba_2Bi_4Ti_5O_{18}$, $Pb_2Bi_4Ti_5O_{18}$, and $Sr_2Bi_4Ti_5O_{18}$ were also prepared using various synthesis routes and their dielectric and ferroelectric responses were evaluated as a function of temperature and frequency. Barium and lead based compounds are found to show relaxor behaviour.

A-site doping in $Ba_2Bi_4Ti_5O_{18}$ and $Sr_2Bi_4Ti_5O_{18}$ thin films is found to improve the ferroelectric behaviour of these compounds with a remarkable increase in the value of remanent polarization. Among all the odd number layered Aurivillius oxides, Pr doped bismuth titanate and Nb doped bismuth titanate thin films can be regarded as potential candidates to replace PZT for commercial applications.

References

[1] Aurivillius B 1952 The structure of Bi_2NbO_5F and isomorphous compounds *Ark. Kemi* **4** 39–47

[2] Mirishli F A and Ismailzade I G 1971 Synthesis and study of new lamellar ferroelectrics *Izv. Akad. Nauk SSSR Ser. Fiz.* **35** 1833

[3] McCabe E E, Jones I P, Zhang D, Hyatt N C and Greaves C 2007 Crystal structure and electrical characterisation of Bi_2NbO_5F: an Aurivillius oxide fluoride *J. Mater. Chem.* **17** 1193–200

[4] Voronkova V I, Kharitonova E P and Rudnitskaya O G 2009 Refinement of Bi_2WO_6 and Bi_2MoO_6 polymorphism *J. Alloys Compd.* **487** 274–9

[5] Newkirk H W, Quadflieg P, Liebertz J and Kockel A 1972 Growth, crystallography and dielectric properties of Bi_2WO_6 *Ferroelectrics* **4** 51–5

[6] Wolfe R W, Newnahm R E and Kay M I 1969 Crystal structure of Bi_2WO_6 *Solid State Commun.* **7** 1797–801

[7] McDowell N A, Knight K S and Lightfoot P 2006 Unusual high-temperature structural behaviour in ferroelectric Bi_2WO_6 *Chem.—Eur. J.* **12** 1493–9

[8] Zeng T, Yan H, Ning H, Zeng J and Reece M J 2009 Piezoelectric and ferroelectric properties of bismuth tungstate ceramics fabricated by spark plasma sintering *J. Am. Ceram. Soc.* **92** 3108–10

[9] Ishikawa K, Watanabe T and Funakubo H 2001 Preparation of Bi_2WO_6 thin films by metalorganic chemical vapor deposition and their electrical properties *Thin Solid Films* **392** 128–33

[10] Xiao-Juan W, Zhi-Qiang G, Jun Z and Xiao-Bing C 2009 Structure, ferroelectric and dielectric properties of Bi_2WO_6 with different bismuth content *Chin. Phys.* B **18** 803

[11] Ahn Y and Son J Y 2016 Ferroelectric properties of highly *a*-oriented polycrystalline Bi_2WO_6 thin films grown on glass substrates *AIP Adv.* **6** 105008

[12] Ahn Y and Son J Y 2017 Ferroelectric properties of highly *c*-oriented epitaxial Bi_2WO_6 thin films *J. Cryst. Growth* **462** 41–4

[13] Van den Elzen A F and Rieck G D 1973 Redetermination of the structure of Bi_2MoO_6, koechlinite *Acta Crystallogr.* B **29** 2436–8

[14] Mączka M, Freire P T C, Luz-Lima C, Paraguassu W, Hanuza J and Mendes Filho J 2009 Pressure-induced phase transitions in ferroelectric Bi_2MoO_6—a Raman scattering study *J. Phys. Condens. Matter* **22** 015901

[15] Buttrey D J, Vogt T and White B D 2000 High-temperature incommensurate-to-commensurate phase transition in the Bi_2MoO_6 catalyst *J. Solid State Chem.* **155** 206–15

[16] Aurivillius B 1951 Mixed oxides with layer lattices. 3. Structure of $BaBi_4Ti_4O_{15}$ *Ark. Kemi* **2** 519–27

[17] Idink H, Srikanth V, White W B and Subbarao E C 1994 Raman study of low temperature phase transitions in bismuth titanate, $Bi_4Ti_3O_{12}$ *J. Appl. Phys.* **76** 1819–23

[18] Lupeiko T G, Lopatin S S and Lisutina L 1986 Preparation of bismuth titanate in the presence of a salt melt of sodium and potassium chlorides *Izv. Akad. Nauk SSSR, Neorg. Mater.* **22** 801–6

[19] Jovalekic C and Stevic S 1992 A study of ferroelectric properties of $Bi_4Ti_3O_{12}$ ceramics prepared from chemically derived powders *Ferroelectrics* **132** 185–96

[20] Shulman H S, Testorf M, Damjanovic D and Setter N 1996 Microstructure, electrical conductivity, and piezoelectric properties of bismuth titanate *J. Am. Ceram. Soc.* **79** 3124–8

[21] Subbarao E C 1961 Ferroelectricity in $Bi_4Ti_3O_{12}$ and its solid solutions *Phys. Rev.* **122** 804

[22] Rae A D, Thompson J G, Withers R L and Willis A C 1990 Structure refinement of commensurately modulated bismuth titanate, $Bi_4Ti_3O_{12}$ *Acta Crystallogr.* B **46** 474–87

[23] Fang P H, Robbins C R and Aurivillius B 1962 Ferroelectricity in the compound $Bi_4Ti_3O_{12}$ *Phys. Rev.* **126** 892

[24] Seok Choi W and Nyung Lee H 2012 Band gap tuning in ferroelectric $Bi_4Ti_3O_{12}$ by alloying with $LaTMO_3$ (TM = Ti, V, Cr, Mn, Co, Ni, and Al) *Appl. Phys. Lett.* **100** 132903

[25] Park B H, Kang B S, Bu S D, Noh T W, Lee J and Jo W 1999 Lanthanum-substituted bismuth titanate for use in non-volatile memories *Nature* **401** 682–4

[26] Maffei N and Krupanidhi S B 1992 Excimer laser-ablated bismuth titanate thin films *Appl. Phys. Lett.* **60** 781–3

[27] Dorrian J F, Newnham R E, Smith D K and Kay M I 1972 Crystal structure of $Bi_4Ti_3O_{12}$ *Ferroelectrics* **3** 17–27

[28] Newnham R E, Wolfe R W and Dorrian J F 1971 Structural basis of ferroelectricity in the bismuth titanate family *Mater. Res. Bull.* **6** 1029–39

[29] Noguchi Y, Miwa I, Goshima Y and Miyayama M 2000 Defect control for large remanent polarization in bismuth titanate ferroelectrics—doping effect of higher-valent cations *Jpn. J. Appl. Phys.* **39** L1259

[30] Hervoches C H and Lightfoot P 1999 A variable-temperature powder neutron diffraction study of ferroelectric $Bi_4Ti_3O_{12}$ *Chem. Mater.* **11** 3359–64

[31] Cummins S E and Cross L E 1968 Electrical and optical properties of ferroelectric $Bi_4Ti_3O_{12}$ single crystals *J. Appl. Phys.* **39** 2268–74

[32] Aurivillius B 1949 Mixed bismuth oxides with layer lattices. II. Structure of $Bi_4Ti_3O_{12}$ *Ark. Kemi* **1** 463–80

[33] Shimakawa Y, Kubo Y, Tauchi Y, Asano H, Kamiyama T, Izumi F and Hiroi Z 2001 Crystal and electronic structures of $Bi_{4-x}La_xTi_3O_{12}$ ferroelectric materials *Appl. Phys. Lett.* **79** 2791–3

[34] Kim S K, Miyayama M and Yanagida H 1996 Electrical anisotropy and a plausible explanation for dielectric anomaly of $Bi_4Ti_3O_{12}$ single crystal *Mater. Res. Bull.* **31** 121–31

[35] Chakraborty K R, Achary S N, Patwe S J, Krishna P S R, Shinde A B and Tyagi A K 2007 Low temperature neutron diffraction studies on $Bi_4Ti_3O_{12}$ *Ceram. Int.* **33** 601–4

[36] Mei X A, Chen M, Liu R F, Sun Y H and Liu J 2012 Negative resistance behavior of ferroelectric bismuth titanate ceramics at low field *Key Eng. Mater.* **492** 234–7

[37] Madeswaran S, Giridharan N V and Jayavel R 2003 Sol–gel synthesis and property studies of layered perovskite bismuth titanate thin films *Mater. Chem. Phys.* **80** 23–8

[38] Wang H, Ren M and Okuyama M 2007 Synthesis and characteristics of nature-superlattice-structured Bi_3TiNbO_9–$Bi_4Ti_3O_{12}$ ferroelectric thin films by MOD *Mater. Sci. Eng.* B **136** 1–4

[39] Xiang P H, Kinemuchi Y, Nagaoka T and Watari K 2005 Sintering behaviors of bismuth titanate synthesized by a coprecipitation method *Mater. Lett.* **59** 3590–4

[40] Liu W, Wang X, Tian D, Xiao C, Wei Z and Chen S 2010 Chemical reaction and crystalline procedure of bismuth titanate nanoparticles derived by metalorganic decomposition technique *Mater. Sci. Appl.* **1** 91

[41] Ng C Y and Razak K A 2011 Properties of praseodymium-doped bismuth potassium titanate ($Bi_{0.5}K_{0.5}TiO_3$) synthesised using the soft combustion technique *J. Alloys Compd.* **509** 942–7

[42] Li W, Chen K, Yao Y, Zhu J and Wang Y 2004 Correlation among oxygen vacancies in bismuth titanate ferroelectric ceramics *Appl. Phys. Lett.* **85** 4717–9

[43] Watcharapasorn A, Siriprapa P and Jiansirisomboon S 2010 Grain growth behavior in bismuth titanate-based ceramics *J. Eur. Ceram. Soc.* **30** 87–93

[44] Kong L B, Ma J, Zhu W and Tan O K 2001 Preparation of $Bi_4Ti_3O_{12}$ ceramics via a high-energy ball milling process *Mater. Lett.* **51** 108–14

[45] Stojanovic B D, Paiva-Santos C O, Jovalekic C, Simoes A Z, Lazarevic Z and Varela J A 2006 Mechanically activating formation of layered structured bismuth titanate *Mater. Chem. Phys.* **96** 471–6

[46] Stojanovic B D, Simoes A Z, Paiva-Santos C O, Quinelato C, Longo E and Varela J A 2006 Effect of processing route on the phase formation and properties of $Bi_4Ti_3O_{12}$ ceramics *Ceram. Int.* **32** 707–12

[47] Han K and Ko T 2009 Synthesis and phase formation of ferroelectric $Bi_4Ti_3O_{12}$ nano-powder by mechanical alloying *J. Alloys Compd.* **473** 490–5

[48] Zdujić M, Jovalekić Č, Poleti D, Veljković I and Karanović L 2009 Mechanochemically induced amorphous/crystalline phase transition in the $Bi_4Ti_3O_{12}$ compound *Mater. Lett.* **63** 2542–4

[49] Du X, Xu Y, Ma H, Wang J and Li X 2007 Synthesis and characterization of bismuth titanate by an aqueous sol–gel method *J. Am. Ceram. Soc.* **90** 1382–5

[50] Du X, Xu Y, Ma H, Wang J and Li X 2008 Low-temperature synthesis of bismuth titanate by an aqueous sol–gel method *J. Am. Ceram. Soc.* **91** 2079–82

[51] Pookmanee P, Uriwilast P and Phanichpant S 2004 Hydrothermal synthesis of fine bismuth titanate powders *Ceram. Int.* **30** 1913–5

[52] Shi Y, Cao C and Feng S 2000 Hydrothermal synthesis and characterization of $Bi_4Ti_3O_{12}$ *Mater. Lett.* **46** 270–3

[53] Xie L, Ma J, Wu P, Tian H, Zhao Z, Zhou J, Hu Y, Wang Y, Tao J and Zhu X 2007 Hydrothermal synthesis and characterization of bismuth titanate regular tetrahedron crystallites *Mater. Res. Bull.* **42** 389–93

[54] Kan Y, Wang P, Li Y, Cheng Y B and Yan D 2002 Low-temperature sintering of $Bi_4Ti_3O_{12}$ derived from a co-precipitation method *Mater. Lett.* **56** 910–4

[55] Subohi O, Kumar G S, Malik M M and Kurchania R 2012 Dielectric properties of bismuth titanate ($Bi_4Ti_3O_{12}$) synthesized using solution combustion route *Physica* B **407** 3813–7

[56] Subohi O, Shastri L, Kumar G S, Malik M M and Kurchania R 2014 Study of Maxwell–Wagner (M–W) relaxation behavior and hysteresis observed in bismuth titanate layered structure obtained by solution combustion synthesis using dextrose as fuel *Mater. Res. Bull.* **49** 651–6

[57] Suresh K and Patil K C 1995 A recipe for an instant synthesis of fine particle oxide materials *Perspectives in Solid State Chemistry* ed J K Rao (New Delhi: Narosa) pp 376–88

[58] Lakshmi V V, Bauri R and Paul S 2011 Effect of fuel type on microstructure and electrical property of combustion synthesized nanocrystalline scandia stabilized zirconia *Mater. Chem. Phys.* **126** 741–6

[59] Shulman H S 1997 Piezoelectric bismuth titanate ceramics for high temperature applications *Thesis* École polytechnique fédérale de Lausanne

[60] Subohi O, Kumar G S and Malik M M 2013 Optical properties and preparation of bismuth titanate ($Bi_{12}TiO_{20}$) using combustion synthesis technique *Optik—Int. J. Light Electron Optics* **124** 2963–5

[61] Subohi O, Kumar G S, Malik M M and Kurchania R 2015 Study of influence of fuel on dielectric and ferroelectric properties of bismuth titanate ceramics synthesized using solution-based combustion technique *Mater. Res. Express* **2** 036302

[62] Mimani T and Patil K C 2001 Solution combustion synthesis of nanoscale oxides and their composites *Mater. Phys. Mech. (Russia)* **4** 134–7

[63] Jimenez B, Jimenez R, Castro A, Millán P and Pardo L 2001 Dielectric and mechanoelastic relaxations due to point defects in layered bismuth titanate ceramics *J. Phys. Condens. Matter* **13** 7315

[64] Kröger F A and Vink H J 1956 Relations between the concentrations of imperfections in crystalline solids *Solid State Phys.* **3** 307–435

[65] Rout S K, Hussian A, Lee J S, Kim I W and Woo S I 2009 Impedance spectroscopy and morphology of $SrBi_4Ti_4O_{15}$ ceramics prepared by soft chemical method *J. Alloys Compd.* **477** 706–11

[66] Mamatha B, Suresh M B, James A R, Vithal M and Sarah P 2011 Synthesis and electrical properties of $SrBi_4Ti_4O_{15}$ piezoelectric ceramics *Phys. Scr.* **84** 055704

[67] Devi C S, Suresh M B, Kumar G S and Prasad G 2012 Synthesis, characterization and electrical properties of Nd/Zr co-doped nano $BaTiO_3$ ceramics *J. Adv. Dielectrics* **2** 1250001

[68] Sharma G D, Roy M and Roy M S 2003 Charge conduction mechanism and photovoltaic properties of 1, 2-diazoamino diphenyl ethane (DDE) based Schottky device *Mater. Sci. Eng.* B **104** 15–25

[69] Prasad K, Chandra K P, Bhagat S, Choudhary S N and Kulkarni A R 2010 Structural and electrical properties of lead-free perovskite $Ba(Al_{1/2}Nb_{1/2})O_3$ *J. Am. Ceram. Soc.* **93** 190–6

[70] Li W, Chen A, Lu X and Zhu J 2005 Collective domain-wall pinning of oxygen vacancies in bismuth titanate ceramics *J. Appl. Phys.* **98** 024109

[71] Lazarevic Z, Stojanovic B D and Varela J A 2005 An approach to analyzing synthesis, structure and properties of bismuth titanate ceramics *Sci. Sinter.* **37** 199–216

[72] Khang T D and Buchanan R C 1997 Orientation effects on pyroelectric properties of solution-derived bismuth titanate films *MRS Online Proc. Libr. Arch.* **493** 509–15

[73] Brody P S 1981 Photovoltaic and other outputs from pulsed laser illuminated ferroelectrics *Ferroelectrics* **38** 939–42

[74] Zhang S T, Cheng H W, Chen Y F, Song C H, Wang J, Xia Y D, Zhao X N, Liu Z G and Ming N B 2004 Composition-dependent structures and properties of $Bi_4Ti_{3-x}Zr_xO_{12}$ thin films *Solid State Commun.* **130** 235–9

[75] Takenaka T, Shoji K, Takai H and Sakata K 1976 Ferroelectric and dielectric properties of press forged $Bi_4Ti_3O_{12}$ ceramics *Proc. Jpn. Congr. Mater. Res* **19** 230–3

[76] Hur S G, Park D H, Kim T W and Hwang S J 2004 Evolution of the chemical bonding nature of ferroelectric bismuth titanate upon cation substitution *Appl. Phys. Lett.* **85** 4130–2

[77] Subohi O, Singh R, Kumar G S, Malik M M and Kurchania R 2015 Impedance analysis and dielectric properties of Ce modified bismuth titanate lead free ceramics synthesized using solution combustion route *J. Mater. Sci., Mater. Electron.* **26** 9122–33

[78] Tang Q Y, Kan Y M, Li Y G, Zhang G J and Wang P L 2006 Effect of vanadium doping on fabrication and property of $Bi_4Ti_3O_{12}$ ceramics *Scr. Mater.* **54** 2075–80

[79] Pavlović N, Koval V, Dusza J and Srdić V V 2011 Effect of Ce and La substitution on dielectric properties of bismuth titanate ceramics *Ceram. Int.* **37** 487–92

[80] Subohi O, Kumar G S, Malik M M and Kurchania R 2016 Effect of B-site isovalent doping on electrical and ferroelectric properties of lead-free bismuth titanate ceramics *J. Phys. Chem. Solids* **93** 91–9

[81] Subohi O, Kumar G S, Malik M M and Kurchania R 2015 Study of electrical and ferroelectric properties of $Bi_{3.4}Ce_{0.6}Ti_{2.4}Zr_{0.6}O_{12}$ ceramics *J. Mater. Sci., Mater. Electron.* **26** 9342–9

[82] Almond D P and Bowen C R 2004 Anomalous power law dispersions in ac conductivity and permittivity shown to be characteristics of microstructural electrical networks *Phys. Rev. Lett.* **92** 157601

[83] Rehrig P W, Park S E, Trolier-McKinstry S, Messing G L, Jones B and Shrout T R 1999 Piezoelectric properties of zirconium-doped barium titanate single crystals grown by templated grain growth *J. Appl. Phys.* **86** 1657–61

[84] Wang L, Gui M, Jin H B, Hu X, Zhao Y, Adnan N M and Li J B 2018 Temperature dependent conductivity of $Bi_4Ti_3O_{12}$ ceramics induced by Sr dopants *J. Adv. Ceramics* **7** 256–65

[85] Tellier J, Boullay P, Jennet D B and Mercurio D 2007 Structure versus relaxor properties in Aurivillius type compounds *J. Eur. Ceram. Soc.* **27** 3687–90

[86] Subohi O, Bowen C R, Malik M M and Kurchania R 2016 Dielectric spectroscopy and ferroelectric properties of magnesium modified bismuth titanate ceramics *J. Alloys Compd.* **688** 27–36

[87] Tripathy M, Mani R and Gopalakrishnan J 2007 New substitutions and novel derivatives of the Aurivillius phases, $Bi_5TiNbWO_{15}$ and $Bi_4Ti_3O_{12}$ *Mater. Res. Bull.* **42** 950–60

[88] Villegas M, Jardiel T and Farıas G 2004 Sintering and electrical properties of $Bi_4Ti_{2.95}W_xO_{11.9+3x}$ piezoelectric ceramics *J. Eur. Ceram. Soc.* **24** 1025–9

[89] Villegas M, Jardiel T, Caballero A C and Fernandez J F 2004 Electrical properties of bismuth titanate-based ceramics with secondary phases *J. Electroceram.* **13** 543–8

[90] Pineda-Flores J L, Chavira E, Reyes-Gasga J, Gonzalez A M and Huanosta-Tera A 2003 Synthesis and dielectric characteristics of the layered structure $Bi_{4-x}R_xTi_3O_{12}$ (R_x = Pr, Nd, Gd, Dy) *J. Eur. Ceram. Soc.* **23** 839–50

[91] Goh P Y, Razak K A and Sreekantan S 2009 Structural and morphology studies of praseodymium-doped bismuth titanate prepared using a wet chemical route *J. Alloys Compd.* **475** 758–61

[92] Yamada M, Iizawa N, Yamaguchi T, Sakamoto W, Kikuta K, Yogo T, Hayashi T and Hirano S I 2003 Processing and properties of rare earth ion-doped bismuth titanate thin films by chemical solution deposition method *Jpn. J. Appl. Phys.* **42** 5222

[93] Kojima T, Sakai T, Watanabe T, Funakubo H, Saito K and Osada M 2002 Large remanent polarization of $(Bi,Nd)_4Ti_3O_{12}$ epitaxial thin films grown by metalorganic chemical vapor deposition *Appl. Phys. Lett.* **80** 2746–8

[94] Shulman H S, Damjanovic D and Setter N 2000 Niobium doping and dielectric anomalies in bismuth titanate *J. Am. Ceram. Soc.* **83** 528–32

[95] Macedo Z S, Ferrari C R and Hernandes A C 2004 Impedance spectroscopy of $Bi_4Ti_3O_{12}$ ceramic produced by self-propagating high-temperature synthesis technique *J. Eur. Ceram. Soc.* **24** 2567–74

[96] Fouskova A and Cross L E 1970 Dielectric properties of bismuth titanate *J. Appl. Phys.* **41** 2834–8
[97] Jimenez B, Jimenez R, Castro A, Millán P and Pardo L 2001 Dielectric and mechanoelastic relaxations due to point defects in layered bismuth titanate ceramics *J. Phys. Condens. Matter* **13** 7315
[98] Noguchi Y, Miwa I, Goshima Y and Miyayama M 2000 Defect control for large remanent polarization in bismuth titanate ferroelectrics—doping effect of higher-valent cations *Jpn. J. Appl. Phys.* **39** L1259
[99] Takahashi M, Noguchi Y and Miyayama M 2003 Effects of V-doping on mixed conduction properties of bismuth titanate single crystals *Jpn. J. Appl. Phys.* **42** 6222
[100] Tang Y *et al* 2018 Enhanced pyroelectric and piezoelectric responses in W/Mn-codoped $Bi_4Ti_3O_{12}$ Aurivillius ceramics *J. Eur. Ceram. Soc.* **38** 5348–53
[101] Yao Y Y, Song C H, Bao P, Su D, Lu X M, Zhu J S and Wang Y N 2004 Doping effect on the dielectric property in bismuth titanate *J. Appl. Phys.* **95** 3126–30
[102] Li W, Gu J, Song C, Su D and Zhu J 2005 B-site doping effect on ferroelectric property of bismuth titanate ceramic *J. Appl. Phys.* **98** 114104
[103] Tang Q Y, Kan Y M, Li Y G, Zhang G J and Wang P L 2007 Ferroelectric and dielectric properties of Nd/V co-doped $Bi_4Ti_3O_{12}$ ceramics *Solid State Commun.* **142** 1–5
[104] Tripathy M, Mani R and Gopalakrishnan J 2007 New substitutions and novel derivatives of the Aurivillius phases, $Bi_5TiNbWO_{15}$ and $Bi_4Ti_3O_{12}$ *Mater. Res. Bull.* **42** 950–60
[105] Subbarao E C 1962 Crystal chemistry of mixed bismuth oxides with layer-type structure *J. Am. Ceram. Soc.* **45** 166–9
[106] Aurivillius B and Fang P H 1962 Ferroelectricity in the compound $Ba_2Bi_4Ti_5O_{18}$ *Phys. Rev.* **126** 893
[107] Onodera A, Mouri S, Fukunaga M, Hiramatsu S, Takesada M and Yamashita H 2006 Phase transition in Bi-layered oxides with five perovskite layers *Jpn. J. Appl. Phys.* **45** 9125
[108] Yi I S and Miyayama M 1998 Electrical properties of layer-structured $Pb_2Bi_4Ti_5O_{18}$ single crystals *J. Ceram. Soc. Jpn.* **106** 285–9
[109] Irie H, Miyayama M and Kudo T 2001 Structure dependence of ferroelectric properties of bismuth layer-structured ferroelectric single crystals *J. Appl. Phys.* **90** 4089–94
[110] Miranda C, Costa M E V, Avdeev M, Kholkin A L and Baptista J L 2001 Relaxor properties of Ba-based layered perovskites *J. Eur. Ceram. Soc.* **21** 1303–6
[111] Kumar S and Varma K B R 2011 Dielectric relaxation in bismuth layer-structured $Ba_2Bi_4Ti_5O_{18}$ ferroelectric ceramics *Curr. Appl Phys.* **11** 203–10
[112] Fukunaga M, Takesada M and Onodera A 2016 Ferroelectricity in layered perovskites as a model of ultra-thin films *World J. Condens. Matter Phys.* **6** 224–43
[113] Fang P H, Robbins C R and Aurivillius B 1962 Ferroelectricity in the compound $Bi_4Ti_3O_{12}$ *Phys. Rev.* **126** 892
[114] Ikegami S and Ueda I 1974 Piezoelectricity in ceramics of ferroelectric bismuth compound with layer structure *Jpn. J. Appl. Phys.* **13** 1572
[115] Subbarao E C 1962 A family of ferroelectric bismuth compounds *J. Phys. Chem. Solids* **23** 665–76
[116] Zou G, Liu J, Cui Q and Yang H 1994 Soft mode phase transition of $Sr_2Bi_4Ti_5O_{18}$ *Phys. Lett.* A **189** 257–60
[117] Dubey S, Subohi O and Kurchania R 2017 Optimization of calcination and sintering temperature of sol–gel synthesised $Ba_2Bi_4Ti_5O_{18}$, $Pb_2Bi_4Ti_5O_{18}$ and $Sr_2Bi_4Ti_5O_{18}$ ceramics *Ceram. Int.* **43** 12755–63

[118] Dubey S, Subohi O and Kurchania R 2018 A comparative study of the properties of five-layered Aurivillius oxides $A_2Bi_4Ti_5O_{18}$ (A = Ba, Pb, and Sr) synthesized by different wet chemical routes *Appl. Phys.* A **124** 1–12
[119] Dubey S and Kurchania R 2017 Effect of sintering temperature on morphology, dielectric and ferroelectric properties of five-layer Aurivillius oxides: $A_2Bi_4Ti_5O_{18}$ (A = Ba, Pb and Sr) synthesized by co-precipitation route *J. Mater. Sci., Mater. Electron.* **28** 8266–77
[120] Kumar S and Varma K B R 2009 Influence of lanthanum doping on the dielectric, ferroelectric and relaxor behaviour of barium bismuth titanate ceramics *J. Phys.* D **42** 075405
[121] Hou R Z and Chen X M 2005 Synthesis and dielectric properties of layer-structured compounds $A_{n-3}Bi_4Ti_nO_{3n+3}$ (A= Ba, Sr, Ca) with $n > 4$ *J. Mater. Res.* **20** 2354–9
[122] Ferrer P, Iglesias J E, Ayala A P, Guedes I and Castro A 2005 Study of the Aurivillius phases $Bi_4Sr_{n-3}Ti_nO_{3n+3}$ (n = 4, 5) synthesized by mechanochemical activation *Solid State Commun.* **136** 621–6
[123] Miyayama M and Yi I S 2000 Electrical anisotropy in single crystals of Bi-layer structured ferroelectrics *Ceram. Int.* **26** 529–33
[124] Yi I S and Miyayama M 1998 Electrical properties of layer-structured $Pb_2Bi_4Ti_5O_{18}$ single crystals *J. Ceram. Soc. Jpn.* **106** 285–9
[125] Irie H, Miyayama M and Kudo T 2000 Electrical properties of a bismuth layer-structured $Ba_2Bi_4Ti_5O_{18}$ single crystal *J. Am. Ceram. Soc.* **83** 2699–704
[126] Shet T, Bhimireddi R and Varma K B R 2016 Grain size-dependent dielectric, piezoelectric and ferroelectric properties of $Sr_2Bi_4Ti_5O_{18}$ ceramics *J. Mater. Sci.* **51** 9253–66
[127] Noumura Y, Sato S, Hiruma Y, Nagata H and Takenaka T 2011 High-power piezoelectric characteristics of $Sr_2Bi_4Ti_5O_{18}$–$Ca_2Bi_4Ti_5O_{18}$-based ferroelectric ceramics *Key Eng. Mater.* **485** 93–6
[128] Cao J, Koval V, Zhang H, Lin Y, Wu J, Meng N, Li Y, Li Z, Zhang H and Yan H 2019 Crystal structure and electrical properties of textured $Ba_2Bi_4Ti_5O_{18}$ ceramics *J. Eur. Ceram. Soc.* **39** 1042–9
[129] Zhang S T, Xiao C S, Fang A A, Yang B, Sun B, Chen Y F, Liu Z G and Ming N B 2000 Ferroelectric properties of $Sr_2Bi_4Ti_5O_{18}$ thin films *Appl. Phys. Lett.* **76** 3112–4
[130] Fang H, Sun H, Zhu J, Mao X Y, He J H and Chen X B 2006 Excellent ferroelectric properties of stoichiometric and Sr-deficient $Sr_2Bi_4Ti_5O_{18}$ thin films prepared by sol–gel process *Integr. Ferroelectr.* **84** 107–14
[131] Song D and Yang J 2019 Lead-free $Ba_2Bi_4Ti_5O_{18}$ thin film capacitors for energy storage applications *Results Phys.* **13** 102175
[132] Irie H, Miyayama M and Kudo T 2000 Electrical properties of a bismuth layer-structured $Ba_2Bi_4Ti_5O_{18}$ single crystal *J. Am. Ceram. Soc.* **83** 2699–704
[133] Wang W, Chen X B, Mao X Y and Fan H X 2006 Properties of Ba-doped $Sr_2Bi_4Ti_5O_{18}$ ferroelectric ceramics *Integr. Ferroelectr.* **85** 103–10
[134] Chen Q, Xu Z, Chu R and Liu Y 2011 Dielectric and ferroelectric properties of (Li, Ce)-doped $Sr_2Bi_4Ti_5O_{18}$ lead-free ceramics *J. Adv. Dielectr.* **1** 439–45
[135] Chen Q, Xu Z J, Chu R Q, Liu Y, Chen M L, Wang X H, Ran C J and Li G R 2011 Study on dielectric and ferroelectric properties of Gd-doped $Sr_2Bi_4Ti_5O_{18}$ ceramics *Adv. Mater. Res.* **328** 1131–4
[136] Qiang F, Lu W P, Zhu J and Chen X B 2004 Dielectric and ferroelectric properties of Dy-doped $Sr_2Bi_4Ti_5O_{18}$ ceramics *Integr. Ferroelectr.* **65** 89–96

[137] Cui L and Hu Y J 2009 Ferroelectric properties of neodymium-doped $Sr_2Bi_4Ti_5O_{18}$ thin film prepared by sol–gel route *Physica* B **404** 150–3

[138] Zhao X, Zhang F, Zhang H, Tian Q, Wang L, Guo X, Shen P, Liu H and Fan S 2019 Effect of Ca doping on the properties of $Sr_2Bi_4Ti_5O_{18}$ ferroelectric thin films *J. Mater. Sci., Mater. Electron.* **30** 13434–44

Chapter 3

Even number layered Aurivillius oxides

This chapter deals with the study of the synthesis and characterization of even number layered (m = 2, 4) Aurivillius oxide ferroelectrics. Different synthesis methods use different processing parameters, such as calcination temperature, sintering temperature, dwell time, and different precursors. These processing parameters affect the phase formation, microstructure, and physical properties of the compounds directly. Hence the synthesis route adopted to prepare these compounds plays a vital role in tailoring the properties of layered oxides. In addition, ion substitution in the host lattice also has a strong influence on its ferroelectric, dielectric, and electrical properties. Variation of these properties as a function of temperature and frequency is discussed in detail.

3.1 Two-layered oxide (m = 2) structure

Layered bismuth (Bi) oxides such as $BaBi_2Nb_2O_9$ (BBN) and $SrBi_2Nb_2O_9$ (SBN), along with their tantalum analogs $BaBi_2Nb_2O_9$ and $SrBi_2Ta_2O_9$, all of which belong to the two-layered Aurivillius oxide family, have received renewed interest following the observation that their ferroelectric properties are superior for applications in memory devices, compared to those of the currently employed $PbZrO_3$–$PbTiO_3$ solid solution (PZT) ceramics [1].

The crystal structures of $BaBi_2Nb_2O_9$ and $BaBi_2Nb_2O_9$ are described as comprising $(Bi_2O_2)^{2+}$ layers and perovskite-like $[BaM_2O_7]^{2-}$ units that contain two layers of corner sharing MO_6 moieties as shown in figure 3.1. It has been suggested using both powder neutron and synchrotron x-ray methods that both BBN and BBT are tetragonal at room temperature, with their structures described by the centrosymmetric space group *I4/mmm*. In spite of this, both BBN and BBT show ferroelectricity at room temperature. The cause of the ferroelectricity in BBN and BBT has been studied widely [2, 3]. In one of the analyses, tetragonal $BaBi_2Nb_2O_9$ is found to comprise bunches with orthorhombic symmetry, and the observed ferroelectric properties are the result of polarization variations in the *a*–*b* plane induced by these domains [2].

Both compositions show relaxor behaviour in ferroelectricity. The relaxor nature is characterized by a strong frequency-dependent broadened dielectric peak [3]. The relaxor behaviour is related to the microscopic distortion in the tetragonal lattice [4] and to the positional disorder between the Bi^{3+} and A^{2+} cations [5, 6]. The existence of cationic disorder was first suggested by Smolenski *et al* [7] in $BaBi_2Ta_2O_9$ and later it was confirmed by neutron diffraction studies. Macquart *et al* [8] used high-resolution synchrotron x-ray-diffraction methods to study the structure of BBN over a wide temperature range. Their investigation confirmed that a crystallographic phase transition takes place near 300 K, which is around the observed ferroelectric to paraelectric transition. They concluded that this phase transition comprises of a change from a low temperature non-centrosymmetric structure belonging to the *I4mm* space group (ferroelectric) to a high temperature centrosymmetric structure with the space group *I4/mmm* (paraelectric).

Some other two-layered Aurivillius type compounds, such as Bi_3TiNbO_9 (BTN) and $CaBi_2Nb_2O_9$ (CBN), possess very high Curie temperatures (>900 °C) and are potential candidates for ultra-high temperature applications [9]. Calcium bismuth niobate $CaBi_2Nb_2O_9$ (CBN) is a piezoelectric material with a Curie temperature of 943 °C, but the piezoelectric coefficient d_{33} of CBN ceramics was found to be about 5 pC N^{-1}. With grain orientation techniques, the piezoelectric coefficient d_{33} could be achieved as higher than 10 pC N^{-1} while maintaining the Curie temperature at 900 °C [10].

3.1.1 Synthesis of $BaBi_2Nb_2O_9$ and Bi_3TiNbO_9 ceramics and thin films

The synthesis of two-layered compounds using the conventional solid-state route has been reported widely in the literature [12, 13]. However, the sol–gel synthesis, co-precipitation, and mechanochemical activation methods have also been employed for successful synthesis of these compounds. Synthesis parameters such as the calcination and sintering temperature and time, the precursors used, and the reaction mechanism are different in each synthesis process. These affect the phase formation

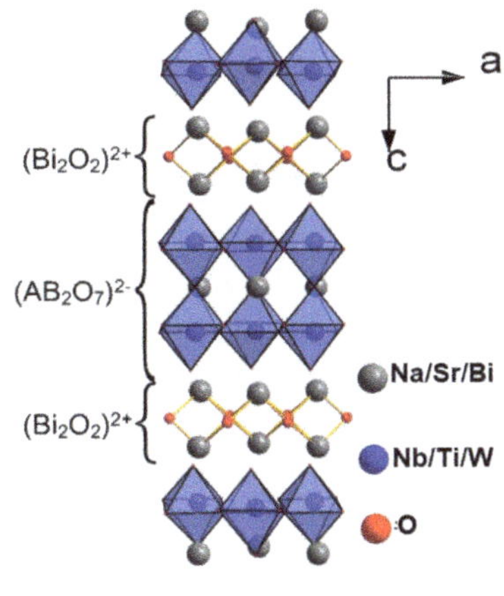

Figure 3.1. BLSF with $m = 2$ layers of pseudo-perovskite. Reproduced from [11] with permission of The Royal Society of Chemistry.

and microstructure of the compounds which, in turn, affect the dielectric, electrical, ferroelectric, and piezoelectric properties of the synthesized material.

Adamczyk *et al* [14] studied the effect of the sintering time on the grain structure and ferroelectric properties of $BaBi_2Nb_2O_9$ ceramics prepared by the solid-state route. They observed that the grain size of the ceramics can be tailored more efficiently by varying the sintering time instead of the sintering temperature. This is because the parameters such as grain size and temperature corresponding to the maximum dielectric constant (T_m) mostly vary linearly with sintering time. However, longer sintering creates defects in the crystal lattice and also significantly reduces the maximum permittivity (ε_{max}). They have also reported two types of dielectric relaxations for BBN ceramics. The first relaxation is observed around the temperature corresponding to the maximum permittivity and obeys the Vogel–Fulcher relationship with a freezing temperature ($T_f = 170$ K) much lower than the temperature corresponding to the maximum permittivity. The other dielectric dispersion at low frequencies occurs in the higher temperature range in the para-electric phase.

The impedance spectroscopy of $BaBi_2Nb_2O_9$ ceramics was studied in the frequency range from 5 Hz to 13 MHz and the temperature range from room temperature (RT) to 500 °C [15]. The impedance spectroscopic data indicate a non-Debye type relaxation mechanism in BBN ceramics. The impedance data were fitted using an equivalent circuit model to estimate the grain and grain boundary resistance at different temperatures. The activation energy due to conduction was calculated using Arrhenius plots and was found to be comparable to that obtained from the values of grain and grain boundary resistivity. These results indicate that in these ceramics the current flows to the same degree at the grain boundaries and grain interiors at higher temperatures. Atomic force microscopy (AFM) with a conductive tip was used as an electrical nanoprobe to measure the conductivity in the contact mode. In this method the small dc voltage applied between the conductive top and the bottom electrode was used for local grain and grain boundary conductivity measurements in contact mode for different tip locations. It was reported that the macroscopic results obtained by impedance spectroscopy were in agreement with those obtained by microscopic direct local LC-AFM conductivity measurements. The impedance values obtained were of the same order of magnitude as that using the microscopic LC-AFM method.

Impedance analysis showed that BBN ceramics possess (i) bulk effects, (ii) a semiconducting nature (negative temperature coefficient of resistance (NTCR)-type behaviour), and (iii) non-Debye-type relaxation phenomena. A linear variation of σ_{ac} with temperature shows that the conduction in these ceramics is a thermally activated phenomenon and obeys the Arrhenius equation. However, complex modulus plots show the presence of both the grain and grain boundary effect in the material, whereas impedance analysis indicates the effect of the grain only [16].

Pure phase $BaBi_2Nb_2O_9$ powders could be obtained at much lower temperatures of 400 °C and 450 °C using a novel hydroxide precursor method [17] and an aqueous sol–gel [18] synthesis method, respectively. It was interesting to note that the ceramics prepared by powders obtained by the sol–gel route showed preferential growth in one direction without any special process. The grain size of these ceramics was found to increase with an increase in sintering temperatures. However, the value of the dielectric constant obtained was smaller compared to that of ceramics prepared through the conventional solid-state route. A lower dielectric constant in sol–gel synthesized ceramics may be attributed to a low density in these ceramics.

Barium bismuth niobate ($BaBi_2Nb_2O_9$), two-layered Aurivillius layered structure perovskites, were synthesized from stoichiometric amounts of barium titanate and bismuth titanate obtained via a mechanochemical activation technique [19]. The mechanochemical synthesis was performed in air atmosphere in a planetary ball mill. The powder mixture of barium titanate and bismuth titanate was homogenized for 30 min, pressed into pellets, and then sintered at 1100 °C for 4 h without the pre-calcination step. The reaction mechanism of $BaBi_2Nb_2O_9$ and the preparation characteristics of the BBT ceramic powders were studied using XRD and Raman spectroscopy, and the particle size was analysed using SEM. Pure phase $BaBi_2Nb_2O_9$ ceramic was obtained at 1100 °C for 4 h without a pre-calcination step. The SEM micrographs reveal a plate-like grain formation which is typical for the Aurivillius oxides and globular and multilateral grains. The Ba^{2+} addition in bismuth titanate leads to a change in its microstructure development, particularly changes in the average grain size.

Powder and thin films of two-layered Aurivillius oxide $BaBi_2Nb_2O_9$ were efficaciously synthesized by a metal–organic decomposition route with a stoichiometric precursor solution [20]. Nano-sized grains were obtained for the BBN powder samples calcined at 700 °C in air. Crack-free films of the BBN phase were obtained on both alumina (α-Al_2O_3) and $Pt/TiO_2/SiO_2/(100)Si$ substrates by spin-coating of an air-stable metal–organic precursor solution of BBN. Good crystallinity in the BBN films deposited on (α-Al_2O_3) and $Pt/TiO_2/SiO_2/(100)Si$ substrates was obtained by annealing the films after deposition at 700 °C for 30 min and 750 °C for 30 min, respectively. X-ray diffraction (XRD) and electromagnetic measurement (EM) techniques were used to understand the phase formation in BBN films. It was observed that the films were polycrystalline with nano-sized arbitrarily oriented grains. More investigations were focused on the optimization of the annealing temperature of BBN films deposited on $Pt/TiO_2/SiO_2/(100)Si$ substrates so as to tailor the properties that are dependent on the orientation of the grains in a particular direction.

Mazon *et al* investigated the growth and properties of BBN thin films deposited on $Pt/Ti/SiO_2/Si$ substrates by the RF magnetron sputtering method [21]. The BBN powders were prepared using the Pechini method at a calcination temperature of 800 °C for 2 h. Pure phase formation of BBN powders was confirmed using XRD analysis. These calcined powders were then uniaxially pressed into pellets 2 inches in diameter and sintered at 1000 °C for 1 h. The thin films of thickness 300 nm were deposited using sintered ceramics. The pressure in the sputtering chamber was kept below 10^{-5} mbar before the deposition process and during the deposition the pressure was maintained at 10^{-2} mbar in an argon and oxygen mixed atmosphere with the $Ar{:}O_2$ ratio kept at 50:50. The RF power used during the deposition was 70 W and the distance between the target and the substrate was 80 mm. The $Ar{:}O_2$

mixture ratio used was found to be optimal for evading bismuth losses during deposition. The films were later annealed in oxygen atmosphere at 700 °C for 1 h and 800 °C for 30 min after deposition. The films thus obtained were dense, smooth, and crack-free. The room temperature dielectric constant and dielectric loss of these films were found to be 780 and 0.09 measured at a frequency of 100 kHz.

The leakage current of $BaBi_2Nb_2O_9$ thin films deposited using the Excimer pulsed laser ablation technique was studied in the temperature range RT to 200 °C [22]. The RT current density evaluated from the *I–V* characteristics was found to be 4.02×10^{-9} A cm^{-2} at an applied electric field of 3×10^5 V m^{-1}. The *I–V* characteristics of the films were linear in nature for an electric field lower than 1 MV m^{-1}. Nonlinear current density–voltage (*J–V*) curves were obtained for the electric field above 1 MV m^{-1}. The *I–V* characteristics of BBN thin films can be explained with the help of different conduction mechanisms. The variation of current density with the applied potential difference of BBN thin films was governed by Lampert's theory of space charge limited conduction in an insulator with traps. Three different regions, i.e. ohmic, trap filled limited, and Child's law, were evidently observed in the *J–V* characteristics. The activation energies in the ohmic regions 0.1 V, 0.3 V, and 0.6 V calculated from the Arrhenius plot were found to be 0.46, 0.48, and 0.51 eV, respectively. These energies could bc ascribed to the shallow traps spread near the conduction band edge in the forbidden gap of the materials. The dielectric response of pulsed laser ablated $BaBi_2Nb_2O_9$ thin films was evaluated as a function of frequency and temperature [23]. Both the dielectric permittivity and dielectric loss of BBN thin films displayed 'power law' dependence with frequency, which was investigated by means of the Jonscher's 'universal dielectric response model'. Theoretical fits were exploited to match the experimental results and also to estimate the value of the temperature-dependent parameters such as s and A used in Jonscher's model. The room temperature permittivity (ε') and dielectric loss ($\tan\delta$) of the BBN thin films were found to be 214 and 0.04, respectively, measured at a frequency of 100 kHz. These films showed the second-order dielectric phase transition from the ferroelectric to paraelectric state at a temperature of 220 °C. The nature of the phase transition was confirmed from the temperature dependence of the dielectric constant and spontaneous polarization. The calculated Curie constant for BBN thin films was 4×10^5C.

The effect of bismuth content on the crystal structure, morphology, and electrical properties of barium bismuth niobate thin films was investigated with the help of x-ray diffraction, scanning electron microscopy, atomic force microscopy, and impedance analyser [24]. BBN was prepared using a polymeric precursor decomposition method and a spin coater was used for the deposition of these thin films on the platinum coated silicon substrates ($Pt/Ti/SiO_2/Si$). The spin coater was operated at 3000 rpm for 30 sec for coating the BBN film. After coating each layer, the films were heated on a hot plate at 90 °C in air for the removal of water and the excess 1.2-ethanediol. These steps were repeated until 300 nm thick film was obtained. A two step procedure was adopted for annealing these films in an oxygen atmosphere. In the first step the films were annealed at 300 °C at a heating rate of 2 °C min^{-1} for removal of organic solvents and in the second stage they were heated up to 700 °C

–800 °C at a heating rate of 5 °C min^{-1} for phase crystallization. The purpose of this two-stage annealing was to obtain crack-free films. The effect of bismuth content on the phase formation, grain size, and morphology of the thin films was also studied. It was observed that single-phase BBN films were obtained for the sample containing additional bismuth up to 2 wt. %. Improved dielectric properties due to larger grain size were observed in the films prepared with excess bismuth content. The BBN thin film with 2 wt. % excess of bismuth possesses a room temperature dielectric constant of around 335 and a dielectric loss of 0.049 at a frequency of 100 kHz.

Another member of the two-layered Aurivillius family Bi_3TiNbO_9 was synthesized using the sol–gel synthesis route. The molar ratio of the deionized water to ethylene glycol was varied to study the effect of this ratio on the particle morphology and crystallization. It was observed that pure phase BTN can be obtained at 600 °C for a water to ethylene glycol molar ratio of 1:0.5 with distinct spherical nanocrystalline particles [25].

Tape casting and hot-pressing technology can be used to produce highly textured and dense Bi_3TiNbO_9 high temperature piezo-ceramics. The preferred orientation of the basal planes and plate-like grains across the pressing axis or along the film casting direction was observed in the ceramics that were prepared by the tape casting and hot-pressing techniques. The piezoelectric constants d_{33} of the as-prepared textured ceramics in directions perpendicular and parallel to the textured plane were found to be 22 pC N^{-1} and 5 pC N^{-1}, respectively [26].

Zhang *et al* [27] synthesized BTNO ceramics using a solid-state reaction and studied the grain orientation effects on the properties of BTNO. They reported that techniques such as hot-press sintering can produce a preferred grain orientation, which was confirmed by SEM. Further, electrical measurements demonstrated that dielectric, piezoelectric, and ferroelectric properties and electrical resistivity show significant anisotropy in hot-pressed samples. In another study [9] it is reported that dielectric plots follow the Curie–Weiss law and the second-order phase transition was observed in even-layered BLSF. The Curie temperature was reported as 914 °C. Microstructural studies showed anisometric plate-like crystalline grains in these ceramics. The electrical resistivity was found to be greater than 10^{11} Ω cm at temperatures below 150 °C and an exponential decrease in resistivity with an increase in temperature was observed. The activation energy due to conduction was found to be 1.13 eV and saturated ferroelectric hysteresis loops could not be observed due to the presence of an internal electric field, which is developed as a result of the asymmetric distribution of charged oxygen vacancies.

3.1.2 The effect of ion substitution in $BaBi_2Nb_2O_9$ and Bi_3TiNbO_9

Ferroelectric, dielectric, and piezoelectric properties can also be tailored by doping the base compound with different ions at the A-site and B-site. The ionic radius, valence state, and position of the dopant ions play vital roles in altering the structure, morphology, dielectric properties, electrical conductivity, and polarization processes of the ceramics.

The effect of vanadium substitution in BBN at the B-site on its impedance behaviour was studied as a function of frequency and temperature [28]. From the impedance spectroscopic data, the grain and grain boundary resistivity were evaluated at different temperatures. The linear variation of these resistivities with temperature can be attributed to the activation nature of these conductivities in both the microstructural regions. Vanadium addition in BBN increases its grain and grain boundary resistivity and also its activation energy, implying a decrease in conductivity of the V doped BBN ceramics. This may be ascribed to the difference in ionic radii of vanadium (V ~ 0.58 Å) and niobium (Nb ~ 0.64 Å). The addition of vanadium tends to make the crystal structure of BBN ordered, which in turn reduces the space charge concentration and conductivity.

Lanthanum substitution in BBN ceramics at the A-site tends to make the grain size distribution more homogeneous, thus decreasing its grain boundary resistivity. The dielectric permittivity of La doped ceramics is found to be lower due to the creation of oxygen vacancies as trivalent La^{3+} ions occupy divalent Ba^{2+} sites. The activation energy is also found to decrease significantly [29].

The effect of rare-earth substitution in barium bismuth niobate was also reported. Cerium and gadolinium doped BBN ceramics with composition $BaBi_{1.8}Ln_{0.2}Nb_2O_9$ (Ln = Ce, Gd) were prepared using the conventional solid-state reaction method [30]. The presence of the parent phase $BaBi_2Nb_2O_9$ was confirmed using the x-ray diffraction pattern. The unit cell parameters were evaluated and the change in the lattice parameters can be correlated with the difference in the ionic radii of the dopants and the host ions. The appearance of Nb–O vibration bands in FTIR spectra is characteristic of barium bismuth niobate phase formation. Dielectric measurements revealed a decrease in the Curie temperature from 340 °C for BBN to 330 °C and 310 °C for BBCeN and BBGdN, respectively. The dielectric constant and loss tangent of pure $BaBi_2Nb_2O_9$ are higher than those of Ce and Gd doped $BaBi_2Nb_2O_9$ ceramics. Moreover, the reduction in the dielectric loss tanδ may be attributed to suppression of oxygen vacancies by the cationic vacancies created by cerium and gadolinium doping.

Cu doped BBN nano ceramics $Ba_{0.5}Cu_{0.5}Bi_2Nb_2O_9$ (BCuBN) were prepared using a chemical precursor solution decomposition method [31]. Structure-dependent dielectric, electrical transport, and ferroelectric properties were studied after the integration of Jahn–Teller active Cu^{2+} ions. XRD and SEM analyses established the orthorhombic structure with the $A2_1am$ space group of BCuBN. The ferroelectric to paraelectric phase transition temperature (T_c) of the as-prepared Cu doped BBN ceramic was found to increase compared to that reported for pure BBN ceramics. It was found to be in the temperature range 515 °C–545 °C at different frequencies. The activation energy of relaxation is 0.80 eV, which is closer to that of dc conductivity (0.73 eV). The frequency-dependent ac conductivity increases drastically at higher frequencies. The *P–E* hysteresis loops validated the pseudo-static saturation of polarization. The maximum discharge energy density was found to be 1.12 J cm^{-3} at a 10 kV cm^{-1} electric field.

The effect of calcium doping on the microstructure, dielectric, and electric properties of $BaBi_2Nb_2O_9$ ceramics prepared using a solid-state reaction and two step sintering process was studied by Szalbot *et al* [32]. Their results show that Ca addition in BBN does not affect the microstructure, such as the shape and size of grains, but it influences its electrical and dielectric properties significantly. A small amount of calcium (1 at %) addition increases the dielectric constant of BBN in the entire temperature range. This increase in the value of ε due to Ca substitution may be attributed to the difference in the ionic radii of the host and the doped ion. In the case of Aurivillius oxides the Bi_2O_2 layers restrict the shrinkage of the unit cell, resulting in increased rattling space for the cations in the octahedral, thus increasing the polarization. Therefore, the dielectric constant of Ca doped BBN increases. However, on further increasing the Ca content to 5 at % in BBN, the dielectric constant decreases. The value of dielectric loss remains almost the same for the doped and undoped BBN from room temperature up to 77 °C and with a further increase in temperature the loss factor increases with an increase in the calcium content. The behaviour of thermally stimulated depolarization currents (TSDC) and impedance spectroscopy indicates the likelihood of the creation of additional oxygen vacancies. These results reveal that the concentration of space charge does not remain constant throughout the sample due to the homovalent nature of the dopant (Ca). In addition, the probability of the creation of additional lattice defects is lower. However, not only the valency of the dopant ions, but also the Ca–O and Ba–O bond strengths should be considered in analysing the defect creation mechanism. As the Ca–O bond is perhaps weaker, loss of the bismuth oxide is expected to increase with an increase in the calcium content. This results in a large number of negative charge carriers connected with unsaturated oxygen ions and the creation of oxygen vacancies. These oxygen vacancies are responsible for a decrease in the resistivity of the ceramics.

A series of polycrystalline sodium doped barium bismuth niobate ceramics $BaBi_{2-x}Nb_2O_9$ with x = 0, 0.02, 0.04, 0.06, and 0.08 were prepared using the conventional mixed oxides method. The phase formation in these Na doped BBN ceramics was determined using the x-ray diffraction pattern. Sharp peaks in the diffraction patterns indicated that the samples were completely crystalline. All the peaks of these XRD patterns were indexed according to the JCPDS card No. 01-0861 191 and were shown to possess tetragonal structure with the space group *I4mm*. No extra peaks were observed in the diffraction patterns of Na doped or undoped BBN ceramics indicating that the as-prepared ceramics were single phase with no secondary or intermediate phases present. The crystallite sizes of these as-prepared ceramics were determined using the Debye–Scherrer method and Williamson–Hall plots. It was found that the crystallite size decreases with an increase in Na content. Rietveld refinement of these diffraction patterns was carried out using Fullprof software, while the lattice parameters, inter-layer distance, bond length, and bond angles were determined for various concentrations of sodium doped BBN ceramics. It was observed that the addition of sodium does not affect the volume of the unit cell. This may be attributed to the fact that as the Na is substituted at the Bi site in Bi_2O_2 layers and the pseudo-perovskite unit ($BaNb_2O_7$) remains unaltered, thus keeping the unit cell volume unchanged. Dielectric spectroscopy of Na doped BBT ceramics was studied in the temperature range RT–450 °C in the frequency range

10 Hz–500 kHz (shown in figure 3.2). Broad maximum curves were obtained for the doped and the undoped samples of BBN, confirming the relaxor nature of these ceramics. It was observed that the temperature (T_m) corresponding to the maximum dielectric constant (ε_m) shifts towards the higher temperature side with an increase in frequency. The diffused phase transition in these BBN ceramics is due to the fact that Ba enters into the Bi_2O_2 layers as well, in addition to the perovskite blocks. This causes an inhomogeneous distribution of Ba ions in the lattice and affects the localized neutrality of the layered structure. This Ba induced disorder in the layered structure is responsible for the diffused phase transition and relaxor nature of the BBN ceramics. Additionally, another dielectric anomaly is observed in the higher temperature region only for lower frequencies. This additional dispersion has been reported previously for perovskite-type ferroelectrics [33]. It arises due to the non-homogeneous distribution of ion space charges participating in the screening process of the polar regions. The occurrence of the polar regions and their interaction with free electron and ion space charges can be identified by thermally stimulated depolarization current (TSDC) studies [33–35]. The degree of dispersion in temperature $\Delta T_m = T_m$ (1 kHz)–T_m (500 kHz) and the degree of dispersion in dielectric constant $\Delta\varepsilon_m = \varepsilon_m$ (1 kHz) − ε_m (500 kHz) decreases with an increase in doping content, indicating a reduction of relaxor behaviour due to Na addition. The presence of Na might be inhibiting the Ba from entering the Bi_2O_2 layers.

The variation of the dielectric constant with temperature for all contents of Na doped BBN ceramics measured at a frequency of 100 kHz is shown in figure 3.3(a). The inset shows the variation of $1/\varepsilon$ with temperature in kelvins. The tangent to the curve determines the value of T_{dev}. In the temperature range between T_m and T_{dev}

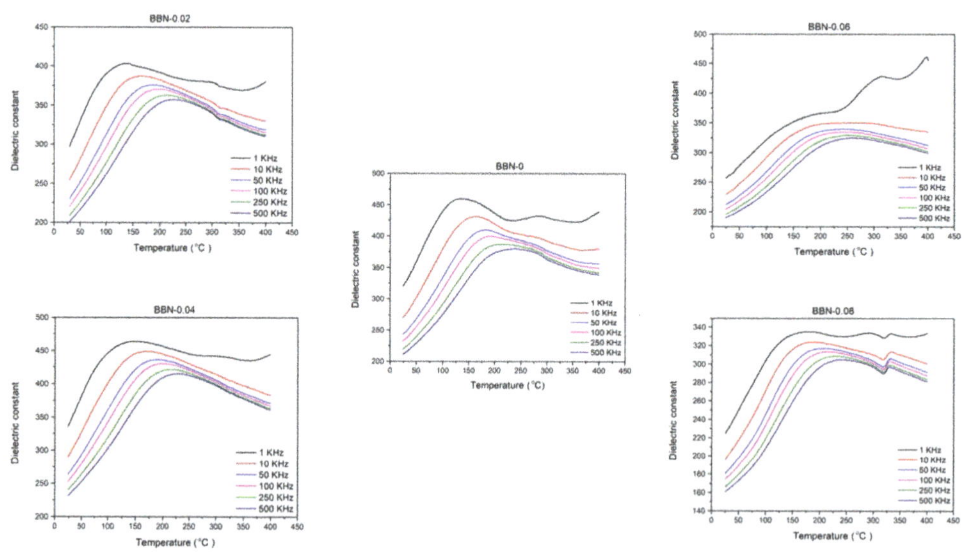

Figure 3.2. Variation of dielectric constant with temperature at various frequencies for $BaBi_{2-x}Nb_2O_9$ ceramics with x = 0, 0.02, 0.04, 0.06, and 0.08.

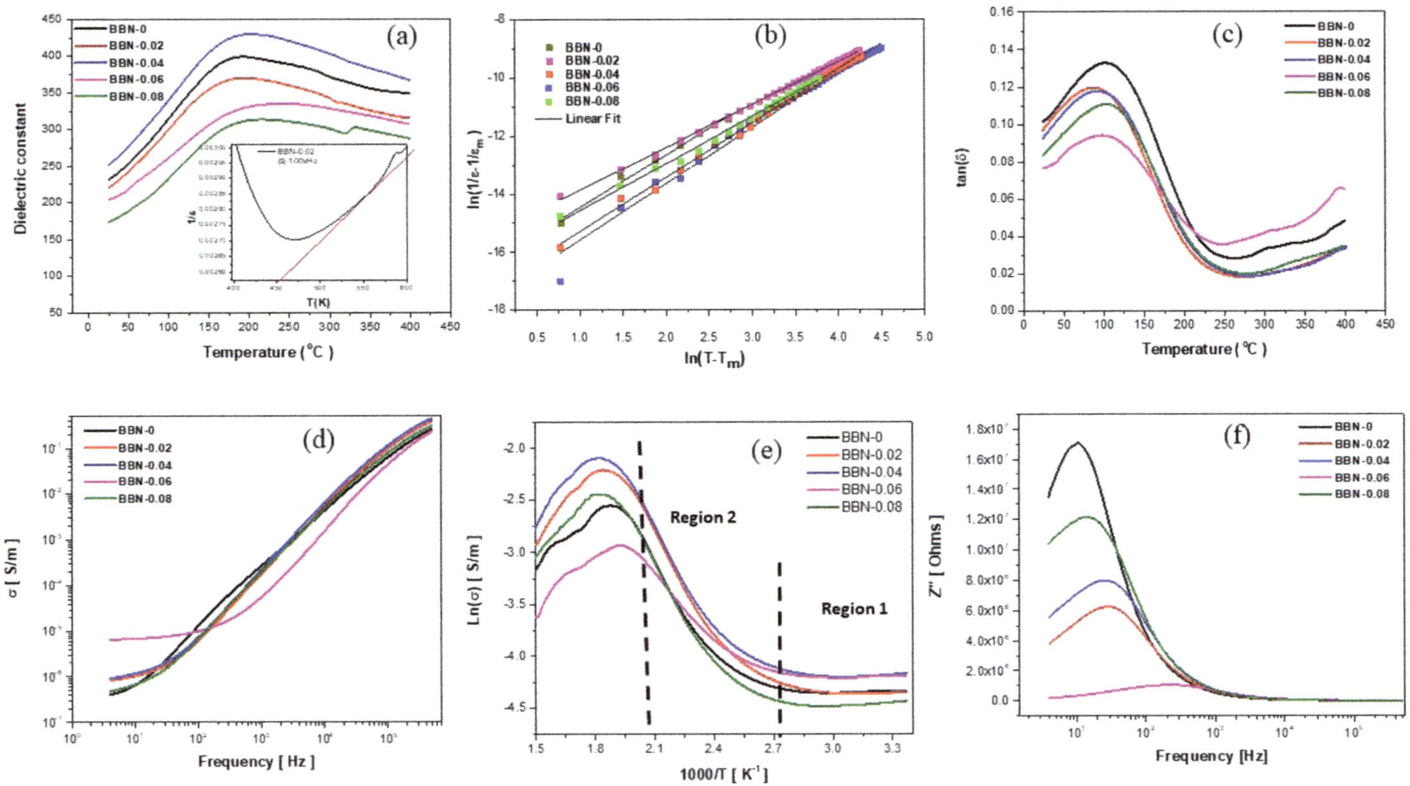

Figure 3.3. Variation of (a) the dielectric constant with temperature, (b) $1/\varepsilon$ with $1000/T$, (c) $\tan\delta$ with temperature, (d) conductivity with frequency, (e) Arrhenius plots at 100 kHz, and (f) imaginary part of impedance with frequency for $BaBi_{2-x}Na_xNb_2O_9$ ceramics.

the function $1/\varepsilon(T)$ could be described by the modified Curie–Weiss law, from which the degree of diffuseness (γ) can be evaluated:

$$\frac{1}{\varepsilon} - \frac{1}{\varepsilon_{\max}} = \frac{(T - T_m)^\gamma}{C}, \tag{3.1}$$

where ε_m is the maximum value of dielectric permittivity at the transition temperature (T_m), C is a constant, and γ is the degree of diffuseness. A graph is plotted between $\ln(1/\varepsilon - 1/\varepsilon_m)$ versus $\ln(T - T_m)$ (shown in figure 3.3(b)) to calculate the value of γ at 100 kHz. It was observed that the value of γ is in the range 1–2 for all samples, which are the limiting values of γ due to which the equation reduces to the Curie–Wiess law, valid for the normal, and the quadratic dependence, valid for the ideal relaxor ferroelectric. In the present case the value of γ increases with an increase in Na concentration up to $x = 0.06$ and then decreases on further increasing Na content to $x = 0.08$. Variation of $\tan\delta$ with temperature for different doping levels of Na is shown in figure 3.3(c). A peak is observed in the $\tan\delta$ versus T curve near T_c. The value of dielectric loss measured at room temperature at a frequency of 100 kHz is found to decrease with an increase in doping level.

The domain structure, domain motion, and the nature of charge carriers present in the system can be studied using conductivity analysis. The influence of bulk conductivity on various properties such as the dielectric and ferroelectric properties and fatigue resistance can be understood by studying the formation mechanism of conducting species in the sample. The classification of the thermally activated conduction mechanism in the materials is as follows:

1. The oxygen vacancies created due to the loss of oxygen during the high temperature sintering process are responsible for the conduction. The temperature- and frequency-dependent conductivity is initiated due to hopping of electrons from one ionic site to another in the sample. The creation and annihilation of oxygen vacancies as a result of their combination with the ions in the material will be responsible for the formation of dipoles that will further influence the dielectric relaxation in the system [36–38].
2. The conduction process in the temperature range near the ferroelectric–paraelectric transition region is due to the creation of small polarons in the crystal as a result of the interaction of the electrons with the polar modes [39].
3. The intrinsic and characteristic conduction mechanism is dominant in the higher temperature region in ferroelectric materials.

The variation of conductivity with frequency for sodium doped barium bismuth niobate ceramics for different concentrations of Na is depicted in figure 3.3(d). The σ versus frequency curves follow Jonsher's power law. There are two regions in the σ versus frequency curve: (i) the frequency-independent region which corresponds to the dc conductivity and (ii) the frequency-dependent region which is designated as ac conductivity. Ionic conductivity due to the movement of doubly ionized oxygen vacancies is responsible for conductivity in these ceramics. It is observed that conductivity increases with an increase in doping level and is found to be maximum for the ceramic with Na = 0.06. The temperature dependence of ac conductivity is shown by Arrhenius plots in figure 3.3(e). It is observed that in the conductivity curves with temperature, two regions exist. In the lower temperature region (Region-1), the activation energy obtained is much lower (0.07–0.1 eV) and in the high temperature region (Region-2) the activation energy due to conduction is found to be in the range (1.9–2.7 eV) for the Na doped BBN ceramics for different values of x. The change in slope from one conduction region to another indicates that there is a change in the conduction mechanism of ions from one temperature regime to another. The low values of the activation energy obtained at lower temperatures may be credited to the hopping charge mechanisms and is linked primarily with the formation of a large number of space charge carriers in the redox process. A hump is observed in the Arrhenius plots of all the samples (doped and undoped BBN) around the Curie temperature which shifts towards higher temperatures with an increase in frequencies.

The electrical transport properties of the materials can be studied as a function of temperature and frequency with the help of impedance spectroscopy [40–42]. It is essential to recognize the role of the grain, grain boundary, and interface effects, which strongly affect the electrical properties. The temperature and frequency-dependent electrical behaviour of the material can be studied by the different formalisms of the complex permittivity (ε^*), complex impedance (Z^*), complex admittance (Y^*), complex electric modulus (M^*), and dielectric loss or dissipation factor ($\tan\delta$), which are related closely to each other [43]. These relationships give insight into a structure-dependent property analysis of different parameters under various physical conditions (temperature or frequency) through graphs. The real and imaginary parts of the impedance, Z' and Z'', respectively, are expressed by the following relations [44]:

$$Z'' = \frac{\omega R}{1 + (\omega\tau)^2} \text{ and } Z' = \frac{R}{1 + (\omega\tau)^2}, \tag{3.2}$$

where $\tau = RC$ is the relaxation time, R and C are the measured resistance and capacitance, and ω is the angular frequency of the electric field.

The variation of the imaginary part of impedance (Z') with frequency at different temperatures for all the doped and undoped ceramics of BBN is evaluated. It is observed that the maximum corresponding to Z'_{max} shifts towards the higher frequency side with an increase in frequency. In addition, the value of the imaginary part of the impedance decreases with an increase in temperature, indicating a negative temperature coefficient of the resistance (NTCR) behaviour of these ceramics in the low frequency region and all these curves merge into a single curve in the higher frequency region depicting frequency-independent behaviour. In addition, the peak corresponding to the maximum broadens with increasing temperature, depicting the multiple temperature-dependent relaxation mechanisms existing in the ceramics. The relaxation process in the ceramics may be due to the heterogenous dispersal of space charge within the grains contributing to the shielding process in the polar regions [45]. The existence of these polar regions

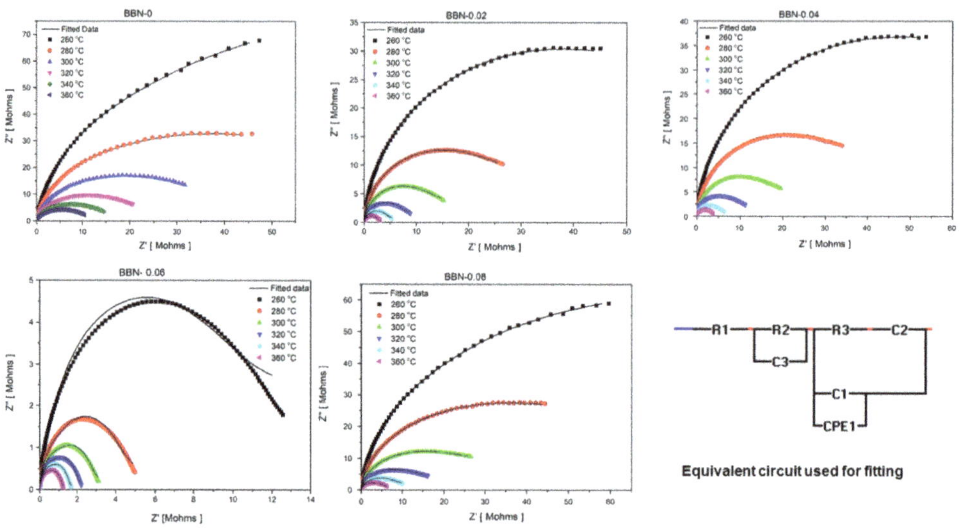

Figure 3.4. The Nyquist plots for $BaBi_{2-x}Na_xNb_2O_9$ for x = 0, 0.02, 0.04, 0.06, and 0.08 and the equivalent circuit used for fitting Cole–Cole plots using EIS software.

and their interaction with free electron and ion space charges may be predicted by thermally stimulated depolarization current (TSDC) analysis. With an increase in the doping content of sodium it is observed from figure 3.3(f) that Z' decreases and is minimum for the sample with $x = 0.06$ and increases on further increasing the doping content up to $x = 0.08$. The value of the relaxation time is evaluated from the frequency corresponding to Z'_{max} at each temperature. The relaxation time is plotted with the inverse of temperature to obtain the values of activation energy due to relaxation.

Figure 3.4 shows the Nyquist plots at different temperatures for all the ceramics corresponding to $BaBi_{2-x}Na_xNb_2O_9$ for x = 0, 0.02, 0.04, 0.06, and 0.08. Nevertheless, the impedance spectroscopic techniques may be employed to extract the information regarding the electrical conduction phenomenon by fitting the experimental impedance data using an equivalent circuit. Typically, the equivalent circuit used for fitting the experimental impedance data consists of resistors and capacitors. However, in certain cases, a constant phase element (CPE) is required to denote non-Debye type behaviour. The CPE depicts the power law dependence of the impedance components on frequency. The CPE impedance is given by

$$Z_{\mathrm{CPE}} = [A_0(j\omega)^p]^{-1}, \tag{3.3}$$

where $A_0 = \frac{A}{\cos(n\pi)/2}$ and $j = \sqrt{} - 1$.

Here, A and p are the frequency-independent but temperature-dependent parameters. The value of p lies between 0 and 1 ($p = 1$ for an ideal capacitor and $p = 0$ for an ideal resistor). The Cole–Cole plots for Na doped BBN ceramics consist of semicircles whose radius lies below the positive x-axis, indicating the existence of a non-Debye type relaxation mechanism. These semicircles can be resolved further

Table 3.1. The values of the grain (R_g) and grain boundary (R_{gb}) resistance obtained from fitted data of Nyquist plots for $Bi_{3-x}Ba_xTiNbO_9$.

Sample →	BBN-0		BBN-0.02		BBN-0.04		BBN-0.06		BBN-0.08	
Temp ↓	R_g (MΩ)	R_{gb} (MΩ)	R_g (MΩ)	R_{gb} (MΩ)	R_g (MΩ)	R_{gb} (MΩ)	R_g (MΩ)	R_{gb} MΩ)	R_g (MΩ)	R_{gb} (MΩ)
260 °C	0.216	0.78	0.182	1	0.25	1.5	0.263	0.338	0.35	1.2
280 °C	0.199	0.61	0.158	0.92	0.150	1.2	0.123	0.212	0.26	0.97
300 °C	0.161	0.42	0.140	0.78	0.138	0.997	0.091	0.166	0.11	0.84
320 °C	0.150	0.35	0.115	0.65	0.121	0.64	0.078	0.138	0.095	0.78
340 °C	0.132	0.32	0.096	0.49	0.117	0.289	0.068	0.103	0.0816	0.15
360 °C	0.120	0.30	0.084	0.218	0.08	0.11	0.045	0.091	0.0755	0.098

into two semicircles—each representing different microstructural regions of the grains and grain boundary. Each microstructural region can be represented by an equivalent parallel circuit of resistance and capacitance. The obtained data of Z' versus Z'' is fitted using the EIS software and the fitted data give the values of the grain and grain boundary resistance shown in table 3.1. It is observed that the grain boundary resistance dominates over grain resistance in all the as-prepared ceramics. In addition, the values of the grain and grain boundary resistance decrease with an increase in temperature for all the ceramics.

Different ion dopings in two-layered Aurivillius oxides such as Bi_3TiNbO_9, $Bi_2CaNb_2O_9$, and $Bi_{2.5}Na_{0.5}Nb_2O_9$ were studied by Bekhtin *et al* [46]. High-density single-phase ceramics of W and Sc/Ta/Ce doped Bi_3TiNbO_9, Nd/Mo and Li/Nd/Sc/W doped $Bi_2CaNb_2O_9$, and Sc doped $Bi_{2.5}Na_{0.5}Nb_2O_9$ were prepared using the hot-pressing technique. The dielectric constant and dielectric loss of the ceramics were measured as functions of temperature and frequency in the temperature range from room temperature to 1000 °C and in the frequency range 25 Hz to 1 MHz. The Curie temperature of the samples was shown to lie in the range T_c = 750 °C–950 °C. A prominent peak in the dielectric constant versus temperature curve and a well-defined minimum in the tanδ versus T curve are observed near T_c at higher frequencies. Even though undoped compounds of Bi_3TiNbO_9, $Bi_2CaNb_2O_9$, and $Bi_{2.5}Na_{0.5}Nb_2O_9$ behave like normal ferroelectrics, their doped counterparts possess a relaxor nature. The maxima and minima in the $\varepsilon(T)$ and tan$\delta(T)$ curves, respectively, tend to shift to the higher temperature side with an increase in frequency. This implies that the relaxation process is associated with the presence of mobile oxygen vacancies in the structure of the phases studied. The resistivity of the samples was measured as a function of temperature and frequency. Above 600 K (327 °C), the resistivity exhibits Arrhenius behaviour with an activation energy E_a = 1.0 eV in the $Bi_3Ti_{0.99}W_{0.01}NbO_9$ sample and 1.6 eV in $Bi_3Ti_{0.96}Sc_{0.02}Ta_{0.02}NbO_9$ + $0.02CeO_2$. The piezoelectric response of the ceramics was studied at room temperature using an alternating load method and the obtained values of piezoelectric coefficient d_{33} lie in the range 11–18 pC N^{-1}.

A series of layered perovskite-like oxides $Bi_{3-x}Gd_xTiNbO_9$ ($x = 0$, 0.1, 0.2, and 0.3) were synthesized and their crystal structure and dielectric properties were studied [47]. The reports suggest that the substitution of the smaller ion Gd^{3+} (1.26 Å for coordination no. 12) in place of Bi^{3+} (1.38 Å for coordination no. 12) in BTNO results in inducing significant distortions in the crystal structure and this influences its dielectric characteristics. Furthermore, the value of the Curie temperature T_c shifts towards the higher temperature side from 921 °C ($x = 0$) to 950 °C (for $x = 0.3$).

Zubkov *et al* [48] prepared yttrium doped BTNO given by the general formula $Bi_{3-x}Y_xTiNbO_9$ ($x = 0$, 0.1, 0.2, and 0.3), and reported that the single-phase compound with orthorhombic unit cells and space group $A2_1am$ was obtained by the solid-state reaction route. They reported an increase in T_c from 921 °C ($x = 0$) to 965 °C (for $x = 0.3$) and the dependence of T_c on the doping concentration x was found to be almost linear.

Single-phase Bi_3TiNbO_9 based ceramics, with Sr^{2+} replacing Bi^{3+} as an acceptor dopant and Nb^{5+} and W^{6+} replacing Ti^{4+} as donor dopants, were prepared using the conventional solid-state reaction method by Zhou *et al* [49]. They reported that conductivity increased with the addition of the acceptor dopant (Sr^{2+}) and decreased due to the donor dopant (W^{6+} and Nb^{5+}), and these results indicate that p-type conduction in pure BTNO is dominant in the low temperature range (200 °C–360 °C). They also evaluated the intrinsic electrical conductivity and activation energy of about 1.8 eV in the high temperature range (365 °C–727 °C).

$Bi_3Ti_{1-x}Nb_{1+x}O_9$ ceramics doped with Nb ions were studied by Zhang *et al* [50] for doping concentrations of ($x = 0$, 0.02, 0.04, 0.06, 0.08, and 0.1). These Nb self-doped BTNO ceramics were synthesized using the conventional solid-state sintering method and it was observed that a higher doping concentration affects the piezoelectric property, ferroelectric property, and resistivity. For $x = 0.04$ the transition temperature (T_c) was found to be 906 °C.

Double ion doping in BTNO at the B-site was studied by Peng *et al* [51]. They reported the crystal structure, and dielectric and piezoelectric properties of Ta and W co-doped BTNO. It was observed that both the Ta and W ions entered into the B-site pseudo-perovskite structure and formed solid solutions. The crystal structure of all the compositions was purely due to the presence of a two-layered Aurivillius type structure without any secondary phases. The Curie temperature for pure BTNO was 896 °C and the T_c had a strong dependence on the structural distortion. Simultaneous modification of the B-site by two ions significantly increases the DC resistivity and piezoelectric properties.

Cerium (Ce) doping in BTNO was found to help in the enhancement of piezoelectric properties. The dielectric properties of Ce doped BTNO showed that the T_c gradually decreases from 907 °C to 889 °C as the cerium content increases to 0.7 wt% [52].

Samarium (Sm) modification in BTNO ($Sm_xBi_{3-x}TiNbO_9$) ceramics affected their crystal structure. The dependence of the lattice parameters with the tolerance factor was studied. A correlation between the lattice distortion induced due to Sm doping in two-layered BTNO ceramics and its dielectric properties was established.

Single-phase formation of $Sm_xBi_{3-x}TiNbO_9$ was confirmed by x-ray diffraction analysis. The tolerance factor (t) was found to increase with an increase in the size of the A-site modified ion (Sm). Dielectric studies show that the amount of dopant or type of dopant does affect the transition temperature [53].

Yan *et al* [54] studied the electrical properties of the BTNO compound as a function of donor and acceptor doping on the B-site. They presented the effect of donor and acceptor doping on the dielectric constant, coercive field, dc conductivity, and piezoelectric constant. These prepared ceramics had a high Curie temperature (~900 °C). They reported that the donor doping results in an increase in ε_r and a decrease in E_c, while the acceptor doping tends to decrease ε_r and increase E_c. They calculated the spontaneous polarization of pure Bi_3NbTiO_9 to be 38.7 $\mu C\ cm^{-2}$ and observed the peak in the dielectric loss curve a few degrees below the Curie temperature, which was attributed to the ferroelectric domain wall movement.

The doping progression in Bi_3TiNbO_9 leads to the origin of various point defects that modifies its physical properties. Among the different dopants, the acceptor dopants, divalent cations such as barium (Ba^{2+}), are of special interest as the polarizability of barium is much greater than that of bismuth. In addition, as the bismuth is highly volatile, when the ceramics are sintered at high temperatures bismuth vacancies are created due to bismuth volatilization, thus degrading the electrical properties due to the increased conductivity in such ceramics. Therefore, the replacement of bismuth ions by barium ions may lower the vacancies, thus improving the ferroelectric properties.

In addition, the valence state of barium ions (Ba^{2+}) is less than that of bismuth ions (Bi^{3+}), so substituting barium in place of bismuth may disturb the electrical neutrality of the ceramic system. Therefore, spontaneous polarization occurs to re-establish the charge neutrality of the compound. The ionic displacement for polarization to occur depends upon the ionic size, tolerance factor, and ionic polarizability. Modification of the host lattice by an acceptor/donor ion at the A-site or B-site changes the A–O or B–O bond lengths, resulting in the tilt of the octahedral, thus affecting the ferroelectric properties of the compound. The tolerance factor of the Ba doped Bi_3TiNbO_9 compounds ($Bi_{3-x}Ba_xTiNbO_9$) was calculated using

$$t = \frac{R_{\mathrm{Aeff}} + R_o}{\sqrt{2}(R_{\mathrm{Beff}} + R_o)}, \tag{3.4}$$

where

$$R_{\mathrm{Aeff}} = \frac{1}{3}[(3 - x)R_{\mathrm{Bi}}^{3+} + xR_{\mathrm{Ba}}^{2+}] \tag{3.5}$$

(R_{Aeff} is the effective radius of cations at the A-site (Bi and Ba)),

$$R_{\mathrm{Beff}} = 0.5[R_{\mathrm{Ti}}^{4+} + R_{\mathrm{Nb}}^{5+}] \tag{3.6}$$

(R_{Beff} is the effective radius of cations at the B-site (Ti and Nb)), and $R_o \rightarrow$ radius of the anion (O).

The equation consists of ionic radii according to the respective coordination number (CN). It is found that the tolerance factor increases with an increase in Ba content. The tolerance factor is 1 for a perfect perovskite compound and is greater than 0.8 for distorted perovskites. In the case of Ba doped Bi_3TiNbO_9 the tolerance factor was found to be between 0.9 and 1.

Polycrystalline ceramics of Ba doped Bi_3TiNbO_9 ($Bi_{3-x}Ba_xTiNbO_9$ for $x = 0$, 0.04, 0.06, and 0.08) were prepared using a solid-state reaction technique. The phase formation of the prepared ceramics was confirmed using x-ray diffraction analysis. The Rietveld refinement of the XRD data is shown in figure 3.5. The variation of the lattice parameters, volume of the unit cell, bond length, bond angles, orthorhombic distortion, and inter-layer distance, as obtained from the refinement data, are plotted as a function of barium concentration and are shown in figure 3.6. It is observed that the volume, orthorhombic distortion, and the inter-layer distance of the unit cell increases with an increase in the doping level of Ba up to $x = 0.06$ and then decreases on further increasing the barium content. Barium (CN = 12) with an ionic radius 1.61 Å, larger than that of bismuth (CN = 12) 1.38 Å, tends to induce lattice distortions, thus varying its lattice parameters. However, beyond the doping level $x = 0.06$ barium might be occupying interstitial positions instead of occupying bismuth vacancies, thus reducing the volume of the unit cell on further increasing the doping level to $x = 0.08$.

Figure 3.7(a) displays a dielectric constant measured at 100 kHz as a function of temperature for $Bi_{3-x}Ba_xTiNbO_9$ ceramics. It is seen that the value of T_c decreases

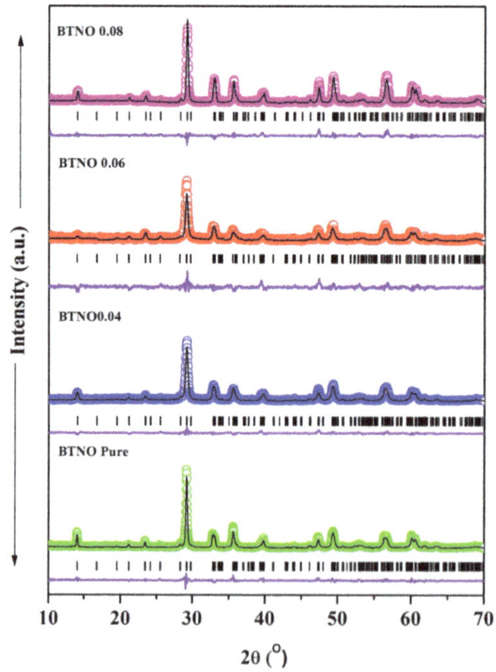

Figure 3.5. Rietveld refinement of the x-ray diffraction pattern of Ba doped Bi_3TiNbO_9.

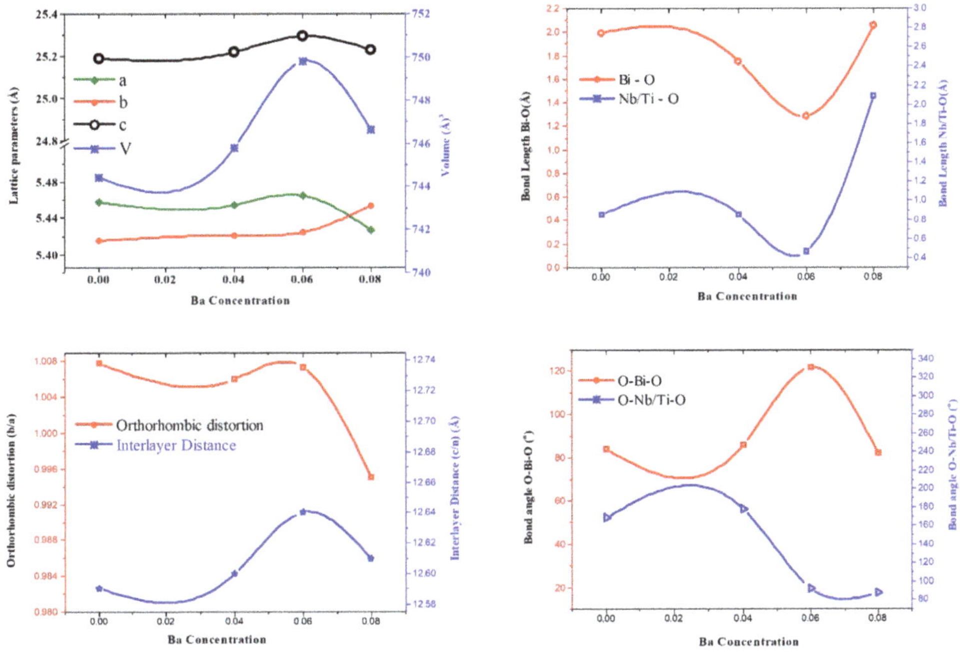

Figure 3.6. The lattice parameters, volume of the unit cell, bond length, bond angles, orthorhombic distortion, and inter-layer distance with barium content.

as the doping concentration of Ba increases from $x = 0.04$ to $x = 0.08$, it has been reported that the Curie temperature increases with decreasing size of the A-site cations in BLSF [55], hence this decrease in T_c must be due to the larger size of the A-site cation (Ba^{2+}) substitution. With increasing doping content, the peak at T_c broadens due to the compositional fluctuation and/or substitution disordering in the arrangement of cations at one or more crystallographic sites in the lattice structure. Figure 3.7(b) presents dielectric loss measured at 100 kHz as a function of temperature for $Bi_{3-x}Ba_xTiNbO_9$. The value of temperature at which the valley occurs in figure 3.7(b) falls around the reported Curie temperature of the respective samples [56]. Above the Curie temperature, there is sharp increase in dielectric loss. Figure 3.7(c) depicts the variation of imaginary part of impedance as a function of frequency for various concentrations of Ba in $Bi_{3\text{-}x}Ba_xTiNbO_9$. Relaxation time was calculated using the frequency corresponding to the maxima of these curves. It was observed that the relaxation time decreases for lower Ba doping concentrations i.e. up to x=0.04 and then increases with further increase in Ba doping. Activation energy due to relaxation was also calculated using relaxation time and was found to increase for lower Ba doping levels from 0.69 eV to 1.34 eV, then decreases on further increase in doping level to 0.65 eV.

The variation of conductivity of the as-prepared ceramics with the inverse of temperature (Arrhenius plots) for $Bi_{3-x}Ba_xTiNbO_9$ ($x = 0$, 0.04, 0.06, and 0.08) are shown in figure 3.7(d). It is observed that with a decrease in temperature the conductivity decreases sharply in the higher temperature range (400 °C–950 °C) and

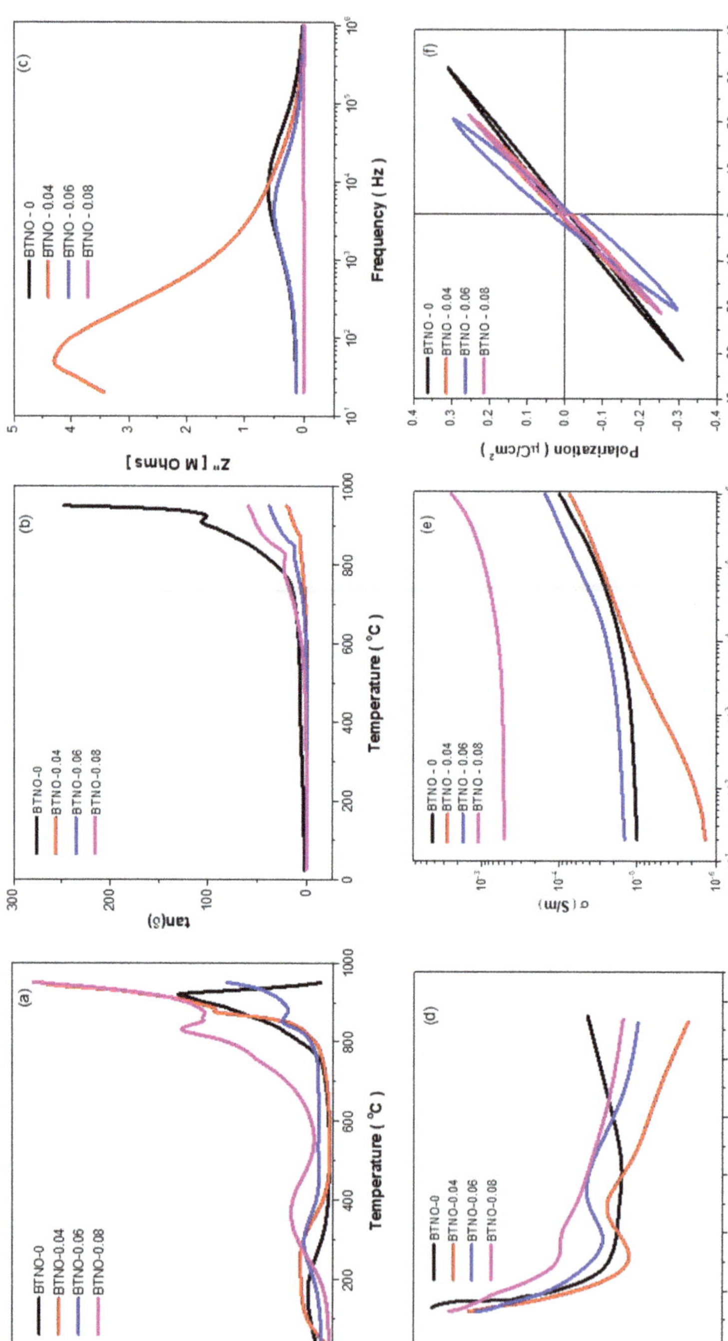

Figure 3.7. Variation of (a) dielectric constant with temperature, (b) dielectric loss with temperature, (c) the imaginary part of the impedance with frequency, (d) Arrhenius plots at 100 kHz, (e) conductivity with frequency at RT, and (f) the P–E loop for $Bi_{3-x}Ba_xTiNbO_9$ ceramics.

Table 3.2. Activation energy due to conduction and relaxation for $Bi_{3-x}Ba_xTiNbO_9$.

Sample name	E_{a1} (eV) due to conduction (400 °C–950 °C)	E_{a2} (eV) due to conduction (RT–300 °C)	$E_{a.}$ (eV) due to relaxation
BTNO-0	1.33	0.03	0.69
BTNO-0.04	2.01	0.08	1.34
BTNO-0.06	1.25	0.07	0.65
BTNO-0.08	1.25	0.21	0.68

almost becomes constant at lower temperatures (RT–300 °C). This is in accordance with the Motts hopping phenomenon [57]. The activation energy due to conduction is calculated using the Arrhenius equation in the higher temperature region (E_{a1}) and lower temperature region (E_{a2}) and is shown in table 3.2. The activation energies due to relaxation are calculated from the frequency (relaxation time) corresponding to the peak in the Nyquist plot. As observed from the conductivity versus frequency graphs, as shown in figure 3.7(e), the plateau region increases with an increase in Ba content, with the hopping frequency shifting towards the higher frequency side. This is indicative of the fact that the conductivity in these samples is due to ionic conductivity. The ferroelectric properties of $Bi_{3-x}Ba_xTiNbO_9$ ceramics were studied using P–E hysteresis loops, as shown in figure 3.7(f). The obtained loops do not show saturation at the applied voltage 30 kV cm^{-1}. The problem of not obtaining saturated hysteresis loops has also been reported previously and domain wall pinning is the reason ascribed for obtaining unsaturated loops. It is observed that remanent polarization increases with an increase in barium content up to $x = 0.06$ and it decreases when the barium content is further increased to $x = 0.08$. This may be ascribed to a decrease in conductivity due to acceptor ion substitution in BTNO up to $x = 0.06$.

The Cole–Cole plots or the Nyquist plots in figure 3.8 for Ba doped BTN show the variation of the imaginary part of the impedance (Z'') with respect to the real part of the impedance (Z'). The plots for all the samples consist of semicircles, which can be resolved further into two semicircles. The centres of these semicircles lie below the Z' axis, which indicates deviation from Debye's relaxation behaviour [58]. The maximum of the semicircle corresponds to the relaxation frequency and the radius of the semicircle gives the resistance. It is observed that with an increase in temperature the radius of the semicircles decreases. This signifies that both the resistance and relaxation time in the as-prepared $Bi_{3-x}Ba_xTiNbO_9$ ceramics decreases with temperature.

These semicircles obtained from the experimental values of Z'' versus Z' are fitted using electrochemical impedance spectroscopy (EIS) software and the parameters associated with the parallel RC element are shown in table 3.3. Figure 3.8 presents the RC circuit used for fitting the semicircles using EIS software. From the values of R_g and R_{gb} it is observed that R_{gb} is greater than R_g, indicating that in the overall electrical behaviour of the prepared ceramics, grain boundary resistance dominates.

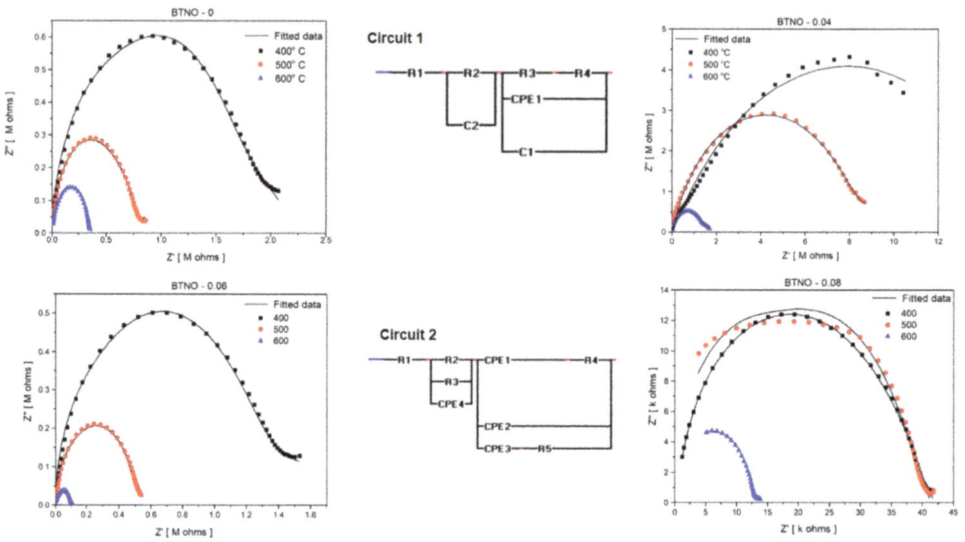

Figure 3.8. Nyquist plots for $Bi_{3-x}Ba_xTiNbO_9$ ($x = 0$, 0.04, 0.06, and 0.08) and the equivalent circuit used for fitting the experimental data. Circuit 1: Used for fitting BTNO-0, BTNO-0.04 at 600 °C, BTNO-0.06, and BTNO-0.08. Circuit 2: Used for fitting BTNO-0.04 at 400 °C and 500 °C.

Table 3.3. The values of the grain (R_g) and grain boundary (R_{gb}) resistance obtained from the fitted data of Nyquist plots for $Bi_{3-x}Ba_xTiNbO_9$.

	BTNO-0		BTNO-0.04		BTNO-0.06		BTNO-0.08	
Temp (°C)	R_{gb} (kΩ)	R_g (kΩ)	R_{gb} (kΩ)	R_g (kΩ)	R_{gb} (kΩ)	R_g (kΩ)	R_{gb} (kΩ)	R_g (kΩ)
400	1830	448.1	1000	947	2000	229	37.1	3.79
500	851	448.1	841	459	491	270	26.68	15.47
600	333	64.5	1870	137	113	18.11	8.41	6.31

The equivalent circuit used for fitting the Nyquist plots consists of two parallel combinations of RC elements—one corresponding to the bulk microstructural region and the other corresponding to the grain boundary.

Therefore, the studies on Ba doped BTN ceramics can be summarized as follows. The barium substitution induces orthorhombic distortion and increases the volume of the unit cell and inter-layer distance up to a doping concentration of $x = 0.06$. The prepared ceramics possess stable dielectric properties and behave like normal ferroelectric materials. The Curie temperature (T_c) decreases as the doping concentration of Ba increases due to the induced lattice distortions. The activation energy due to conduction was calculated from Arrhenius plots and the activation energy due to relaxation was calculated from impedance plots. In all the samples the activation energy due to conduction was found to be greater than that due to relaxation. Impedance analysis suggests that in the overall electrical behaviour of

$Bi_{3-x}Ba_xTiNbO_9$ ceramics, grain boundary resistance dominates. The conductivity in these samples is due to ionic conductivity. Barium doping in Bi_3TiNbO_9 enhances its ferroelectric properties with an increase in remanent polarization up to a doping content of $x = 0.06$. On further increasing the doping content the remanent polarization decreases.

3.2 Four-layered oxides ($m = 4$)

3.2.1 Multiferroic Aurivillius oxide $Bi_5Ti_3FeO_{15}$

Lately, the Aurivillius phase oxides have been reconsidered as one of the potential candidates in single-phase multiferroics because of their magnetoelectric coupling behaviour [59]. $Bi_5Ti_3FeO_{15}$ (BTF) is one such compound with four-layered perovskite units of $(Bi_3Ti_3FeO_{13})^{2-}$ sandwiched between two $(Bi_2O_2)^{2+}$ layers along the *c*-axis. The $(Bi_2O_2)^{2+}$ layers play a key role in both the space charge compensation and insulation, which is expected to reduce the leakage current [60]. The ferroelectric Curie point is reported to be about 750 °C corresponding to a structural transition from *A21am* to *I4/mmm* [61]. However, BTF is antiferromagnetic with $T_N = 80$ K [62]. $Bi_5Ti_3FeO_{15}$ (BTF) may be regarded as a solid solution and/or composite of ferroelectric $Bi_4Ti_3O_{12}$ and $BiFeO_3$ and can be used for different types of devices including electrical transformers, sensors, and electromagnetic memories [63].

The $Bi_5Ti_3FeO_{15}$ ceramics were prepared using a conventional solid-state reaction route possessing orthorhombic structure [64]. The temperature variation of the dielectric constant in BTF ceramics showed no abnormality up to 500 °C. The impedance studies show the existence of a negative temperature coefficient of resistance (NTCR) in the material and dominance of grain resistance. Modulus analysis depicts that the material obeys the non-exponential type of conductivity relaxation. The ac conductivity study follows Jonscher's power law.

Jartych *et al* showed that the processing conditions of three synthesis techniques, i.e. solid-state sintering, mechanical activation, and electrical discharge assisted mechanical milling (EDAMM), have a noteworthy effect on the structure and magnetic properties of $Bi_5Ti_3FeO_{15}$ ceramics [65]. The solid-state sintering method was found to produce pure phase samples. Other methods such as mechanical activation (MA) and EDAMM are also advantageous with respect to the temperature of the heat treatment or the dual time of synthesis, however, the obtained material consists of secondary/intermediate phases. Both these synthesis routes require more research and changes to minimize the secondary phases to a least value. The magnetic properties of $Bi_5Ti_3FeO_{15}$ were studied using Mossbauer spectroscopy. All the as-prepared compounds were in a paramagnetic state at room temperature. Two different sites of Fe^{3+} ions in the crystal lattice were identified and both possess non-cubic environments, hence the quadrupole splitting of the constituent doublets in Mossbauer spectra should be different. The numerical fitting justified the two doublets with similar isomer shifts and different quadrupole splitting in the Mossbauer spectra.

Aurivillius phase multiferroic $Bi_5Ti_3FeO_{15}$ (BTF) polycrystalline thin films were fabricated using a chemical solution deposition route [66]. The morphology, and electrical and magnetic properties, in particular the optical properties of the BTF films, were analysed. The multiferroic behaviour of the BTF films was demonstrated by the coexistence of the ferroelectric and weak ferromagnetic properties at room temperature. The Tauc–Lorentz and Cauchy models were used to fit the experimental data to obtain the optical dielectric functions in the photon energy range of 1.13–4.13 eV. A direct inter-band transition was allowed and the energy bandgap was evaluated to be ~3.3 eV for the BTF films. The Urbach band tail rule described the weak absorption below the bandgap edge. The optical dispersion behaviour of BTF films in the transparent region was elucidated using a single oscillator model.

Silva *et al* synthesized single-phase Aurivillius oxide $Bi_5Ti_3Fe_{0.5}Ni_{0.5}O_{15}$ (BTFN) ceramics using the solid-state reaction method [67]. The replacement of half of the Fe ions by Ni does not introduce a magnetic impurity phase but increases the magnetic moment by more than two orders of magnitude. Curie temperatures corresponding to the ferroelectric and ferromagnetic phases were found to be 830 °C and 450 °C. The room temperature multiferroic behaviour of the BTFN ceramics was confirmed by the ferroelectric *P–E* loop ($2P_r = 8.5\ \mu C\ cm^{-2}$, $2E_c = 74\ kV\ cm^{-1}$) and a ferromagnetic *M–H* curve ($2M_r$ = 27.86 m emu g^{-1}, $2H_c$ = 553 Oe) measurements. The ferromagnetism can be ascribed to the aggregation of magnetic ions at the inner octahedra by Ni doping and the spin canting of magnetic ion-based sub-lattices via the Dzyaloshinskii–Moriya interaction. The work by Silva *et al* outlines the prospects of doped $Bi_5Ti_3FeO_{15}$ as a potential room temperature single-phase multiferroic material.

Aurivillius phase oxide $Bi_3R_2Ti_3FeO_{15}$ (R = Bi, Nd, and Gd) ceramics were synthesized using the traditional solid-state reaction route by Chen *et al* [68]. All the samples possessed an orthorhombic structure with a space group $A2_1am$. The compounds $Bi_3Nd_2Ti_3FeO_{15}$ and $Bi_3Gd_2Ti_3FeO_{15}$ showed reduced orthorhombicity in comparison with $Bi_5Ti_3FeO_{15}$. The temperature variation of magnetic susceptibility obeys the Curie–Weiss law with the Curie–Weiss temperature taking negative values, thus suggesting that the magnetic interactions are antiferromagnetic in nature. Weak canted antiferromagnetic behaviour at temperatures below 25 K and linear behaviour at high temperatures was shown by the magnetization curves. Mossbauer spectroscopy measurements show an increase in the quadrupole splitting with a decrease in temperature, demonstrating the presence of local distortions, favouring the formation of a weak ferromagnetic phase via the antisymmetric Dzyaloshinskii–Moriya interaction in the samples.

$Bi_5Ti_3FeO_{15}$ (BTF15) (m = 4) thin films belonging to the family of Bi-layer-structured multiferroics were prepared on (001) and (110) oriented $SrTiO_3$ (STO) single crystal substrates using pulsed laser deposition by Nakashima *et al* [69]. The x-ray diffraction patterns of these thin films on (001) STO single crystals show that the obtained thin films were (00*l*) oriented layer perovskite single-phase films. The thin films deposited on STO (110) single crystal resulted in (111) oriented layered perovskite films. For studying the ferroelectric properties, the BTF15 thin films were deposited on (001) and (110) oriented La doped (3.73 wt%) STO single crystal substrates. The *P–E* hysteresis loops of BTFO15 ($m = 4$) and BIT-BTFO ($m = 3$–4) thin films on (110) La doped STO single crystals shows good ferroelectric hysteresis loops and their remanent polarization ($2P_r$) was found to be 47 μC cm^{-2} and 44 μC cm^{-2}, respectively.

The effect of holmium (Ho) doping in $Bi_5Ti_3FeO_{15}$ films on its ferroelectric bipolar cyclic properties and magnetoelectric properties was investigated by Bai *et al* [70]. The doped and undoped BTF films were fabricated using the sol–gel method. It was reported that the ferroelectric and magnetoelectric performance of the as-prepared pure BTF films deteriorated after 10^9 bipolar electric cycles. The ferroelectric and leakage fatigue behaviour are ascribed mostly to the defect dipoles and other agglomerated space charges on the surface after bipolar electrical cycling, which further results in magnetoelectric fatigue. Compared to the pure $Bi_5Ti_3FeO_{15}$ films, the Ho doped $Bi_5Ti_3FeO_{15}$ films show much stronger fatigue resistance. The ferroelectric polarization and magnetoelectric coefficient decreased slightly in Ho doped films. The good magnetoelectric fatigue resistance derives mainly from the depression of ferroelectric and leakage fatigue. Therefore, Ho doping in BTF films may be regarded as a tool to tailor the anti-fatigue behaviour of magnetoelectric films for applications in functional devices.

Koval *et al* synthesized the Aurivillius phase $(Bi_{4.3}Gd_{0.7})(Fe_{1-x}Co_x)_{1+y}Ti_{3-2y}Nb_yO_{15}$ ($x = 0$, 0.3, 0.5, and $y = 0$, 0.3) ceramics using a solid-state reaction method to understand the effect of cobalt (Co) doping on ferromagnetic properties in the temperature range above ambient temperature [71]. A single-phase four-layered Aurivillius structure with a space group $A2_1am$ was observed in the cobalt-free configuration, while the cobalt substitution produced an intergrowth structure consisting of four- and three-layered phases. Rietveld refinement of the synchrotron XRD data showed a structural phase transformation from $A2_1am$ to *B2cb* at the morphotropic phase boundary (MPB) of this composition due to the decrease in the number of layers. High-resolution transmission electron microscopy (HRTEM) micrographs show a disordered intergrowth of this composition which may be ascribed to the presence of nanoscale structural modulation at MPB. Mossbauer spectra reveal a sextet which suggests long-range magnetic ordering in the doped samples. The ferromagnetic nature of these Co doped Aurivillius phase ceramics was confirmed from the *M–H* hysteresis loops measured at higher temperatures. The maximum values of the remanent and saturation magnetization at room temperature were obtained for the composition with $x = 0.3$. This may be attributed to the following three reasons: (i) ferromagnetic bunching of the FeO_6 and CoO_6 octahedra, (ii) the spin canting effect, and (iii) the double exchange magnetic interaction between the mixed valence cobalt through oxygen. The cooperative freezing of randomly distributed Fe–O–Co bunches is believed to be accountable for the spin-glass-like behaviour observed at lower temperatures. Therefore, an improvement in the ferromagnetic nature of the ferroelectric cobalt doped Aurivillius ceramics may be due to the nanoscale structural modification (NSM) brought in due to Co substitution and the accumulation of the ferromagnetic bunches near the 3–4 layer interface of the intergrowth.

Palizdar *et al* [72] prepared highly textured, (002) oriented $Bi_5Fe_{0.5}Co_{0.5}Ti_3O_{15}$ (BFCT) samples using molten salt synthesis. However, SEM and XRD analyses reveal the presence of 2 vol% of Co–Fe spinel secondary phases in the as-prepared BFCT, which was much lower than that prepared using the conventional route. The ferromagnetic performance of these Aurivillius oxides was dominated by the presence of secondary phases. A relationship was obtained between the magnitude of the remnant magnetization and the concentration of secondary phases present in the sample. It was inferred that the ferromagnetic ordering in these highly textured ceramics was not due to the multiferroic property of this Aurivillius phase, but it is mostly composition-dependent. This was confirmed by in- and out-of-plane magnetic measurements, which suggests that presence of the Co–Fe spinel phase is responsible for the ferromagnetic performance of the sample. The ferroelectric *P–E* loop measurements showed higher conductivity due to the leakage current in the conventionally prepared BFCT ceramics, compared to those prepared using the molten salt route. However, the ferroelectricity could not be confirmed in either of the ceramics due to a lack of saturation. Ferroelectricity was confirmed for the conventional preparation method via the presence of a butterfly loop.

3.2.2 The four-layered titanates

Strontium bismuth titanate ($SrBi_4Ti_4O_{15}$—SBT15) has been reported to have been prepared using various synthesis routes, such as the conventional solid-state reaction method, mechanochemical activation, and wet chemical routes—the co-precipitation, citrate gel, solution combustion, sol–gel, and hydrothermal methods. Each of these methods uses different precursors with different reaction mechanisms under varying conditions of temperature, pressure, and dual time.

Aurivillius oxide $SrBi_4Ti_4O_{15}$ (SBT15) was synthesized using mechanochemically assisted methods [73]. The phase formation was studied at different stages of sample preparation by x-ray diffraction at room and high temperatures, thermal analysis, and scanning electron microscopy. The effect of different milling media such as vibrating mills (VM) and planetary mills (PM) and the length of the mechanical treatment were studied in understanding the phase formation mechanism. Stainless steel vessels and balls were used for mechanical activation in the vibrating and planetary mills. The sample evolution under mechanical treatment was followed using XRD. It was observed that increasing the milling time (VM or PM) reduces the annealing temperature and time for obtaining pure phase SBT powders compared to those prepared using a solid-state reaction route. The conventional solid-state reaction method requires several intermittent heat treatments for long periods (~12 h) and grinding, and the pure phase is obtained only after heating at 1050 °C. In contrast, the precursor samples from either VM for 168 h and PM for 12 h obtained by mechanical activation produced oxide mixed with the fluorite phase after 1 h at 450 °C, and the pure phase when the precursors were heated at 600 °C for 5 and 1 h, respectively. In the case of the precursor PM for 108 h, the pure Aurivillius phase could be obtained after 1 h heat treatment at 550 °C. Longer heating times and higher temperatures can therefore be dispensed with when starting with a precursor which has been previously mechanically activated.

Four-layered SBT15 was synthesized by a hydrothermal reaction with 20% acetic acid solution as the solvent at 200 °C for 24 h [74]. The obtained pellets were pressed into disks and sintered in air at 1100 °C for 3 h to form ceramic samples. XRD results indicated that well-crystallized SBT was identified at 200 °C without any significant secondary or intermediate phases. The sintered samples were composed of plate-like grains with random orientations. In addition, T_c is higher than 520 °C while the tanδ is low and very stable, even when the measurement temperature was increased to 400 °C. The activation energy (E_a) values of the ionic conductivity suggested that the defects were related to oxygen vacancies. The optimum values of the remnant polarization ($2P_r$) and piezoelectric coefficient (d_{33}) were 13.6 μC cm^{-2} and 19 pC N^{-1}, respectively.

Single-phase $SrBi_4Ti_4O_{15}$ (SBT15) fine powders were obtained from simple salts through co-precipitation and citrate gel techniques [75]. The XRD patterns recorded for both the materials and the particle size measurements on the pure single or Aurivillius phase were carried out using the Scherrer equation. The dielectric constants are reported for the samples prepared by both methods at frequencies of 20–800 kHz

Polycrystalline SBT15 ceramic was prepared using the sol–gel method [76]. XRD pattern analysis showed that the SBT15 ceramic crystallizes in a perovskite-type orthorhombic structure. Plate-like grain morphologies were observed. The experimental density of SBT15 was found to be 7.04 gcm^{-3}, which is 96% that of the theoretical value. The material has been confirmed as single phase with orthorhombic structure, having lattice parameters a = 5.457 Å, b = 5.458 Å, and c = 41.18 Å. These values are similar to the reported values. The Curie temperature (T_c) of SBT was 535 °C, which almost coincides with the reported values. The dielectric constant is 1148 and the loss is 0.03 at T_c; the dielectric loss (tanδ) is very low at RT (about 0.009) and almost constant from RT to 400 °C. The dc conductivity activation energy values are 0.018 and 1.73 eV in the ferro- and paraelectric regions. Oxygen vacancies are considered to be the most mobile charge carriers in layered perovskite ferroelectrics. The evaporation of bismuth oxide in layered perovskite would lead to oxygen vacancies. Impedance analysis indicates the presence of grain and grain boundary effects in SBT15. The magnitude of Z' decreases with an increase in temperature and there is a shift in the peak frequency towards the high-frequency side due to a change in relaxation in the sample, leading to non-Debye type relaxation. The impedance spectroscopy analysis in SBT15 shows that the conduction mechanism is dominated mostly by grain. A clear change in ac conductivity is observed in the high-frequency region below and above the dielectric transition temperature, indicating a change in electrical property of the material around the transition temperature. The electromechanical coupling constant k_p and piezoelectric charge coefficient d_{33} are 0.57 and 16 pC N^{-1}, respectively. These findings make SBT15 a potential material for applications in high temperature piezoelectrics.

Strontium-based layered perovskite is currently one of the most promising candidates for new generation non-volatile ferroelectric random-access memory

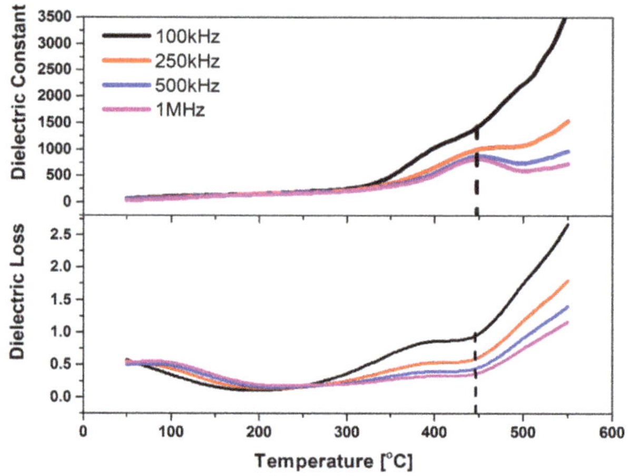

Figure 3.9. the Variation of dielectric constant and dielectric loss as a function of temperature at different frequencies for $SrBi_4Ti_4O_{15}$ synthesized using the solution combustion route.

(NvFRAM) devices. Among several bismuth layer structure ferroelectrics (BLSF) materials, $SrBi_4Ti_4O_{15}$ (SBT) has been extensively studied by many researchers for possible applications in piezoelectric devices. Recently, a lot of attention has been paid to SBT due to its high Curie temperature, low coercive field, barrier type property, large retentivity, and anisotropic physical properties [77]. SBT was prepared using a solution combustion technique with glycine as the fuel. Single-phase formation of the layer-structured compound of SBT with orthorhombic structure was achieved after calcination at 800 °C, and was confirmed by XRD studies. A scanning electron micrograph shows that the grains exhibit a plate-like morphology and possess small grain size. Variation of the dielectric constant and dielectric loss as a function of temperature at various frequencies was studied and is shown in figure 3.9. It was observed that $SrBi_4Ti_4O_{15}$ obtained using the solution combustion technique shows a peak in the ε versus temperature, confirming its normal ferroelectric behaviour. Its ferroelectric to paraelectric transition temperature (T_c) is found to be 450 °C. The values of the dielectric constant decrease with an increase in frequency. Figure 3.9 shows the variation of the dielectric loss (tanδ) with temperature at different frequencies. Lower values of dielectric loss are observed at lower temperatures and it increases sharply beyond T_c. A kink in the tanδ versus temperature curve is observed at T_c, indicating a phase transition from the ferroelectric to paraelectric phase. The ferroelectric properties of SBT obtained by the solution combustion method were studied using a P–E loop, as shown in figure 3.10. A well-saturated P–E loop was obtained depicting the ferroelectric nature of the as-prepared ceramic. Remnant polarization of $2P_r = 1.84\ \mu C\ cm^{-2}$ at a coercive field $2E_c = 2.61$ kV cm^{-1} was obtained.

Aurivillius phase $BaBi_4Ti_4O_{15}$ (BBT15) ceramics were prepared using the conventional mixed oxide route and their electrical transport properties were investigated by Badapanda et al [78]. XRD analysis was used for the phase and structural determination of the as-prepared BBT15 ceramics. It was found that they possessed

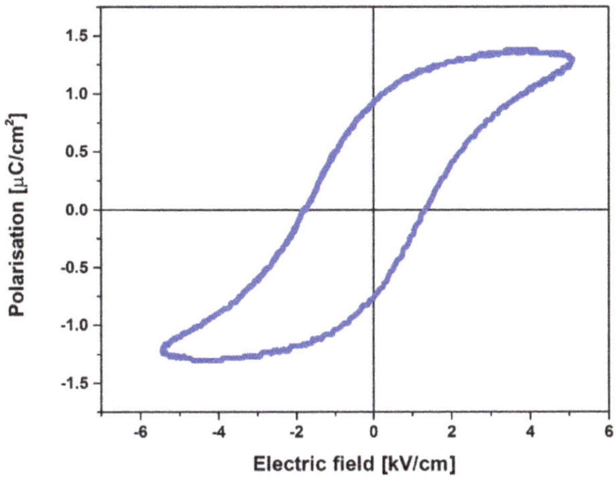

Figure 3.10. Room temperature *P–E* loop for $SrBi_4Ti_4O_{15}$ synthesized using the solution combustion route.

an orthorhombic crystal structure with space group $A2_1am$ corroborating it as a four-layered member of the Aurivillius family. The SEM micrographs of the sintered ceramics depicted grains with a typical plate-like morphology, which is a distinctive feature of the Aurivillius compounds. The relaxor-type behaviour of the as-prepared ceramics was confirmed from the shift of the peak corresponding to ε_{max} with frequency, and a diffuse phase transition in the dielectric constant versus temperature curves. The diffuseness parameter γ was also evaluated. The relaxor behaviour of these ceramics was ascribed to the A-site cationic disorder induced due to presence of Ba at the A-site. The dielectric constant curves followed the Vogel–Fulcher relation and different parameters such as activation energy for relaxation, freezing temperature, and relaxation frequency were evaluated from nonlinear curve fitting. The ε versus T curves were fitted using two different exponential functions in the higher and lower temperature regions, corresponding to ε_{max}, to understand the origin of polar bunches at high temperatures and the dispersal of freezing temperatures in the low temperature region.

Bobic *et al* prepared barium bismuth titanate (BBT15) ceramics using the traditional solid-state route (SSR) and mechanochemical (MC) route to understand the role of the synthesis route on the electrical properties of BBT15 ceramics [79]. The formation of single-phase Aurivillius oxide was established by an x-ray diffraction study of the sintered ceramics prepared by both routes. Low temperature synthesis of the BBT15 powders with smaller particle size was achieved using the mechanochemical synthesis route compared to those from the solid-state technique. SEM micrographs showed the complete grain formation in the ceramics prepared using both routes and the grain size of the BBT15 ceramics obtained using the MC route was also found to be smaller than those obtained using the SSR. From the results of ΔT_{relax} it was confirmed that the synthesis route does not affect the relaxation process as the same amount of structural disorder existed in both ceramics. As reported for other members of the Aurivillius family of compounds, it was observed that the conductivity in the ceramics decreases with a decrease in grain size, which may be due to an increase in the number of grain boundaries. Remanent polarization was found to decrease with a decrease in grain size.

$BaBi_4Ti_4O_{15}$ ceramics belonging to the family of Aurivillius compounds were prepared using the soft chemical route [80]. The precursors were obtained by precipitating bismuth and barium oxalates from a TiO_2 powder suspension. A pure phase BBT15 powder with orthorhombic structure was obtained by heating the precursor powder at 1000°C. TG-DSC and XRD analysis were used to study the phase formation behaviour. Grain formation and the microstructure of the sintered ceramics were studied using SEM micrographs. Variation of the dielectric constant with temperature in the range RT–500 °C and frequency in the range 1 kHz–1 MHz was examined. The diffuse dielectric constant peaks were found to be frequency-dependent. The activation energy, freezing temperature, and relaxation frequency were evaluated from the Vogel–Fulcher relation fitting parameters and were found to be 0.2639 eV, 620 K, and 4.95×10^{21} Hz, respectively. The obtained parameters show that the as-prepared $BaBi_4Ti_4O_{15}$ possesses a relaxor nature.

Photocatalytic properties of Aurivillius phase $BaBi_4Ti_4O_{15}$ powders synthesized using the solid-state reaction were reported by Qi *et al* [81]. X-ray diffraction analysis showed that the as-prepared BBT15 powders possessed orthorhombic structure. The temperature-dependent dielectric behaviour of $BaBi_4Ti_4O_{15}$ ceramics showed broad peaks in the $\varepsilon(T)$ curves, depicting diffuse phase transition around the temperature ~410 °C. Dielectric curves were fitted using the Vogel–Fulcher law and the freezing temperature was determined as 274 °C. The relaxor nature of the ceramics was also confirmed from the current–electric field (*I–E*) loop which showed discrete ferroelectric domain switching current peaks and a room temperature piezoelectric coefficient $d_{33} = 7.0 \pm 0.1$ pC N^{-1} was determined. The direct bandgap (3.2 eV) of the as-prepared BBT15 ceramics was evaluated from UV–vis absorption spectra. The photocatalytic activity of BBT15 powders was evaluated by studying the degradation of rhodamine B (RhB) solution using as-prepared $BaBi_4Ti_4O_{15}$ powders. It was observed that RhB was degraded by 15% after an exposure to UV–vis radiation for 3.5 h in the presence of BBT15 powders. The rate of degradation of RhB was further improved by depositing Ag nanoparticles on the $BaBi_4Ti_4O_{15}$ powder surface. Therefore, BBT15 powders can be regarded as a potential candidate for photocatalytic applications.

$Bi_{4.5}Na_{0.5}Ti_4O_{15}$ is a member of the BLSF family with $m = 4$ (where m is the number of BO_6 octahedrons in the pseudo-perovskite block). The crystal structure of $Bi_{4.5}Na_{0.5}Ti_4O_{15}$ consists of repeated layering of four pseudo-perovskite units alternating with intermediate $(Bi_2O_2)^{2+}$ layers along the *c*-axis [82, 83]. In this, some of the Bi-sites in the pseudo-perovskite block have been occupied by Na atoms. $Bi_{4.5}Na_{0.5}Ti_4O_{15}$ is considered to be a potential candidate for both ferroelectric and piezoelectric applications as it has a high ferroelectric transition temperature T_c, good fatigue resistance, and a high mechanical quality factor Q [84–86]. $Bi_{4.5}Na_{0.5}Ti_4O_{15}$ has an orthorhombic crystal structure which undergoes phase transition at 675 °C.

The synthesis of $Bi_{4.5}Na_{0.5}Ti_4O_{15}$ by various chemical routes has been reported by many researchers. Jannet *et al* [87] reported the synthesis of sodium bismuth titanate $Na_{0.5}Bi_{4.5}Ti_4O_{15}$ (NBT) and barium bismuth titanate $BaBi_4Ti_4O_{15}$ (BBT) ceramics using a solid-state reaction. On increasing the barium content in NBT, the ferroelectric to paraelectric phase transition gradually changes from a normal second-order type to a relaxor-like behaviour above 60 mol.% of barium. The T_c or T_m temperatures decreased from 655 °C (NBT) to 400 °C–420 °C (BBT) as a consequence of the substitution of larger 12-fold coordination Ba^{2+} cations for smaller (Na, Bi^{3+}) cations.

The stearic acid gel method was used for the synthesis of sodium bismuth titanate $Na_{0.5}Bi_{4.5}Ti_4O_{15}$ (NBT) by Supriya *et al* [88]. These NBT powders were sintered at 1200 °C for 1 h. The phase formation was confirmed by XRD and FTIR analysis. The morphology and grain size were observed using SEM analysis. The thermal properties were also analysed for this compound

The synthesis of praseodymium (Pr) doped $Na_{0.5}Bi_{4.5}Ti_4O_{15}$ ($NBT_{-x}Pr^{3+}$) ceramics using a solid-state reaction process was reported by Chen *et al* [89]. $NBT_{-x}Pr^{3+}$ ceramics with $x = 0.3$ were found to possess a maximum dielectric constant of 1109 and a minimum dielectric loss of 0.0082, E_a of 1.122 eV, a piezoelectric constant (d_{33}) = 18 pC N^{-1}, and a well-saturated P–E loop with remanent polarization ($2P_r$) of 12.06 μC cm^{-2}.

A series of $Na_{0.5}Bi_{4.5}Ti_4O_{15}$ -based nominal compositions were studied by Kuai *et al* [90]. The obtained flexoelectric-type polar ceramics contain $Na_{0.5}Bi_{4.5}Ti_4O_{15}$ and $Bi_{12}TiO_{20}$ phases, and their macroscopic polarization might originate from the partial alignment of distorted TiO_6 octahedra and BiO_5 polyhedra that are located within the grain boundary amorphous phases. The piezoelectric coefficient d_{33} was found to be larger than 8 pC N^{-1} and had a depoling temperature higher than 700 °C.

Raghavan *et al* [91] reported the synthesis of pure $Na_{0.5}Bi_{4.5}Ti_4O_{15}$ and rare-earth doped $Na_{0.5}Bi_4RE_{0.5}Ti_4O_{15}$ (RE = Ce, Pr, and Tb) thin films were prepared on Pt (111)/Ti/Si/SiO_2(100) substrates through a chemical solution deposition method. The Tb^{3+} ion leads to a remarkable improvement in the ferroelectric properties. The use of the Tb^{3+} ion for doping gives rise to a well-saturated ferroelectric hysteresis loop. The thin film was found to have a large remnant polarization ($2P_r$) of 40 μC cm^{-2} and a low coercive electric field ($2E_c$) of 176 kV cm^{-1}, measured at an applied electric field of 475 kV in the $Na_{0.5}Bi_4Tb_{0.5}Ti_4O_{15}$ thin film. The leakage current density of the $Na_{0.5}Bi_4Tb_{0.5}Ti_4O_{15}$ thin film was one order of magnitude lower than that of the $Na_{0.5}Bi_{4.5}Ti_4O_{15}$ thin film.

Gao *et al* [92] reported the synthesis of $Na_{0.5}B_{i4.5}Ti_4O_{15}$ + x wt. % $MnCO_3$ by a conventional sintering technique. The results show that $MnCO_3$ doping helps in promoting grain growth and densification of the ceramics. The addition of $MnCO_3$ gives rise to an increase in remanent polarization P_r, piezoelectric coefficient d_{33}, electromechanical coupling coefficient k_p, and mechanical quality factor Q_m of the ceramics, while the loss tangent (tanδ) decreases. For the ceramic doped with 0.375 wt.% $MnCO_3$, the dielectric and piezoelectric properties obtained are $d_{33} = 28$ pC N^{-1}, $k_p = 0.06$, $Q_m = 3290$, $\varepsilon_r = 140$, and tanδ = 0.3%.

Lazarević *et al* [93] reported the synthesis of barium bismuth titanate, $BaBi_4Ti_4O_{15}$ (BBT), using mechanochemical synthesis. The mechanochemical synthesis was performed in air atmosphere in a planetary ball mill. The microstructure of $BaBi_4Ti_4O_{15}$ shows plate-like grains typical for the Bi-layered structured material and spherical and polygonal grains. There were changes in the microstructure development after Ba^{2+} doping, particularly in the change of the average grain size.

Kalaiselvi *et al* [94] reported the synthesis of $SrBi_4Ti_4O_{15}$ by co-precipitation and citrate gel techniques. It is found that $SrBi_4Ti_4O_{15}$ is an orthorhombic structure. The lattice parameters $a = 5.428$, $b = 5.438$, and $c = 40.94$ Å were determined using XRD. The average particle size obtained from the Debye–Scherrer equation using XRD data for the co-precipitation and citrate gel methods were 120 and 140 nm, respectively. The dielectric constants obtained for the $SrBi_4Ti_4O_{15}$ ceramics were 2616 and 2585 measured at 100 kHz and at 520 °C for the co-precipitation and citrate gel methods, respectively.

The synthesis of the four-layered Aurivillius compound $Na_{0.5}B_{i4.5}Ti_4O_{15}$ by the co-precipitation route at different calcination (500 °C, 600 °C, and 700 °C) and sintering temperatures (900 °C and 1000 °C) was carried out. The role of calcination temperature on phase formation and the influence of sintering temperature on the ferroelectric and dielectric properties were investigated [95]. The x-ray diffraction analysis reveals that 700 °C is the optimal calcination temperature at which pure phase $Na_{0.5}B_{i4.5}Ti_4O_{15}$ was obtained. The ferroelectric and dielectric properties were found to improve with an increase in sintering temperature due to the formation of higher density grains. Figure 3.11 shows the *P–E* hysteresis. The ceramics sintered at 1000 °C were found to show a higher value of remanent polarization ($2P_r = 0.38$ μC cm^{-2}), dielectric constant (23.5 at 1 kHz), and low dielectric loss, compared to the sample sintered at 950 °C. However, leakage current was observed in these ceramics due to the creation of bismuth vacancies at higher temperatures.

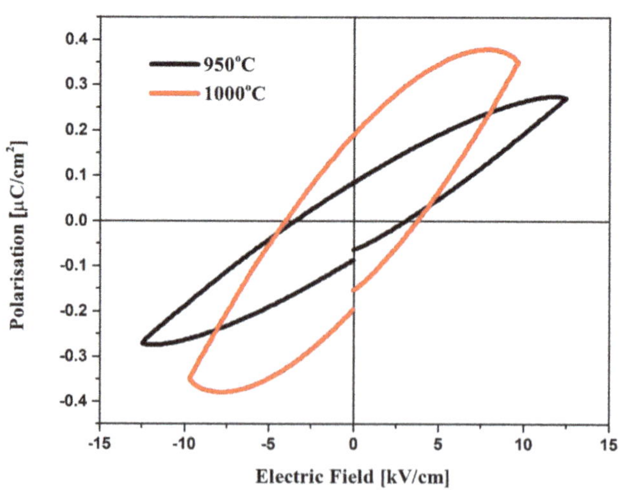

Figure 3.11. *P–E* loop for $Na_{0.5}B_{i4.5}Ti_4O_{15}$ obtained using the co-precipitation route. [95] (2019) Copyright. With permission of Springer.

To summarize, even number layered ($m = 2, 4$) Aurivillius oxide compounds in the form of ceramics and thin films have been studied over a wide frequency and temperature range by various researchers. Two- and four-layered barium-based niobates, tantalates, and titanates all possess relaxor behaviour due to the presence of cationic disorder at the A-site. $BaBi_2Nb_2O_9$ belonging to the two-layered family of Aurivillius oxides possesses tetragonal crystal structure with the space group *I4mm* in the ferroelectric phase, which is non-centrosymmetric and its symmetry changes to the centrosymmetric space group *I4/mmm*. It possesses a relaxor nature with diffused phase transition due to non-homogeneous cation distribution. Bi_3TiNbO_9 is a high temperature normal ferroelectric material with a Curie temperature of ~930 °C. It possesses good ferroelectric properties. It has an orthorhombic crystal structure with the $A2_1am$ space group. A sharp peak is obtained in $\varepsilon(T)$ curves at T_c. These ceramics obtained by the tape casting and hot-pressing technique are found to have a high piezoelectric coefficient of 22 pC N^{-1} along the direction perpendicular to the textured plane. Different types of doping were used to tailor the properties of BBN and BTNO. Rare-earth doping in BBN reduces its Curie temperature with enhancement in its dielectric properties. Cu doped BBN ceramics are found to show an increase in T_c with improved dielectric properties. Acceptor doping of sodium at a site of BBN does not affect the volume of the unit cell but reduces the relaxor nature of these ceramics. Ba doping in BTNO at the bismuth site improves its ferroelectric properties. $Bi_5Ti_3FeO_{15}$ belonging to the four-layered Aurivillius family may be regarded as a single-phase multiferroic compound due to its good magneto-dielectric coupling. Ni doped $Bi_5Ti_3FeO_{15}$ shows room temperature multiferroicity with $2P_r = 8.5\ \mu C\ cm^{-2}$ and $2M_r = 27.86$ m emu g^{-1}. Rare-earth doping enhances the ferroelectric and piezoelectric properties of $Na_{0.5}Bi_{4.5}Ti_4O_{15}$ ceramics and thin films. Tb^{3+} doped $Na_{0.5}Bi_{4.5}Ti_4O_{15}$ thin films show a remarkably high value of remanent polarization ($2P_r$) = 40 $\mu C\ cm^{-2}$ at a low coercive field ($2E_c$) = 176 kV cm^{-1}. $MnCO_3$ addition in $Na_{0.5}B_{i4.5}Ti_4O_{15}$ ceramics significantly improves its piezoelectric properties with $d_{33} = 28$ pC N^{-1}, $k_p = 0.06$, and $Q_m = 3290$.

References

[1] Macquart R, Kennedy B J, Hunter B A, Howard C J and Shimakawa Y 2002 Structural phase transitions in the ferroelectric oxide $SrBi_2Ta_2O_9$ *Integr. Ferroelectr.* **44** 101–12

[2] De Araujo C P, Cuchiaro J D, McMillan L D, Scott M C and Scott J F 1995 Fatigue-free ferroelectric capacitors with platinum electrodes *Nature* **374** 627–9

[3] Blake S M, Falconer M J, McCreedy M and Lightfoot P 1997 Cation disorder in ferroelectric Aurivillius phases of the type $ABi_2Nb_2O_9$ (A= Ba, Sr, Ca) *J. Mater. Chem.* **7** 1609–13

[4] Kennedy B and Ismundar 1999 Effect of temperature on cation disorder in $ABi_2Nb_2O_9$ (A = Sr, Ba) *J. Mater. Chem.* **9** 541–4

[5] Kennedy B J, Howard C J and Chakoumakos B C 1999 Phase transitions in perovskite at elevated temperatures—a powder neutron diffraction study *J. Phys. Condens. Matter* **11** 1479

[6] Shimakawa Y, Kubo Y, Nakagawa Y, Goto S, Kamiyama T, Asano H and Izumi F 2000 Crystal structure and ferroelectric properties of $ABi_2Ta_2O_9$ (A= Ca, Sr, and Ba) *Phys. Rev.* **B 61** 6559

[7] Smolensky G A 1961 New ferroelectrics of complex composition *IV. Sov. Phys.-Solid State* **2** 2651–4

[8] Macquart R, Kennedy B J, Vogt T and Howard C J 2002 Phase transition in $BaBi_2Nb_2O_9$: implications for layered ferroelectrics *Phys. Rev.* B **66** 212102

[9] Zhang Z, Yan H, Dong X and Wang Y 2003 Preparation and electrical properties of bismuth layer-structured ceramic Bi_3TiNbO_9 solid solution *Mater. Res. Bull.* **38** 241–8

[10] Yan H, Zhang H, Ubic R, Reece M J, Liu J, Shen Z and Zhang Z 2005 A lead-free high-Curie-point ferroelectric ceramic, $CaBi_2Nb_2O_9$ *Adv. Mater.* **17** 1261–5

[11] Long C, Fan H, Li M and Li Q 2012 Effect of lanthanum and tungsten co-substitution on the structure and properties of new Aurivillius oxides $Na_{0.5}La_{0.5}Bi_2Nb_{2-x}W_xO_9$ *Cryst. Eng. Commun.* **14** 7201–8

[12] Li B, Wang X and Li L 2003 The sintering behavior of $SrBi_2Ta_2O_9$ associated with synthesis temperature of its powder in solid-state reaction *Mater. Lett.* **57** 1987–91

[13] Peng D, Sun H, Wang X, Zhang J, Tang M and Yao X 2012 Blue excited photoluminescence of Pr doped $CaBi_2Ta_2O_9$ based ferroelectrics *J. Alloys Compd.* **511** 159–62

[14] Adamczyk M and Pawełczyk M 2009 Sintering time dependence of $BaBi_2Nb_2O_9$ ceramics properties *Arch. Metall. Mater.* **54** 979–84

[15] Adamczyk M, Kozielski L and Pilch M 2011 Impedance spectroscopy of $BaBi_2Nb_2O_9$ ceramics *Ferroelectrics* **417** 1–8

[16] Patri S K, Deepti P L, Choudhary R N P and Behera B 2018 Dielectric, impedance and modulus spectroscopy of $BaBi_2Nb_2O_9$ *J. Electroceram.* **40** 338–46

[17] Muthurajan H, Rao N K, Gupta U N, Pradhan S, Jha R K, Kumar H H, Mirji S A and Ravi V 2008 Novel hydroxide precursors for low temperature synthesis of selected ternary oxides *Mater. Res. Bull.* **43** 1842–9

[18] Liu W, Wang H, Zhang B and Li Y 2012 $BaBi_2Nb_2O_9$ powder and their ceramics prepared by aqueous solution–gel method *J. Sol-Gel Sci. Technol.* **64** 673–7

[19] Lazarević Z Ž, Bobić J, Romčević N Ž, Paunović N and Stojanović B D 2009 Study of barium bismuth titanate prepared by mechanochemical synthesis *Sci. Sinter.* **41** 329–35

[20] Benčan A, Boullay P and Mercurio J P 2004 Characterisation of $BaBi_2Nb_2O_9$ powders and thin films prepared by a solution synthesis technique *Solid State Sci.* **6** 547–51

[21] Mazon T, Joanni E, Fernandes J R A, Zaghete M A, Cilense M and Varela J A 2003 Growth and characterization of $BaBi_2Nb_2O_9$ thin films made by RF-magnetron sputtering *Ferroelectrics* **293** 201–7

[22] Laha A and Krupanidhi S B 2002 Leakage current conduction of pulsed excimer laser ablated $BaBi_2Nb_2O_9$ thin films *J. Appl. Phys.* **92** 415–20

[23] Laha A, Krupanidhi S B and Saha S 2002 Dielectric properties of pulsed excimer laser ablated $BaBi_2Nb_2O_9$ thin films *MRS Online Proc. Libr.* **748** 66

[24] Mazon T, Zaghete M A, Cilense M and Varela J A 2009 Effect of the excess of bismuth on the morphology and properties of the $BaBi_2Nb_2O_9$ thin films *Ceram. Int.* **35** 3143–6

[25] Yin H, Zhou A, Chang N and Xu X 2009 Characterization and photocatalytic activity of Bi_3TiNbO_9 nanocrystallines synthesized by sol–gel process *Mater. Res. Bull.* **44** 377–80

[26] Spitsin A I, Bush A A, Kamentsev K E, Segalla A G, Khramtsov A M and Chistyakova N A 2018 Microstructure and electrical transport properties of Bi_3TiNbO_9 high-temperature piezoceramics *Inorg. Mater.* **54** 736–43

[27] Zhang Z, Yan H, Xiang P, Dong X and Wang Y 2004 Grain orientation effects on the properties of a bismuth layer-structured ferroelectric (BLSF) Bi_3TiNbO_9 solid solution *J. Am. Ceram. Soc.* **87** 602–5
[28] Adamczyk M, Kozielski L, Bochenek D, Radoszewska D, Pawełczyk M and Wodecka-Duś B 2016 Impedance spectroscopy of vanadium modified $BaBi_2Nb_2O_9$ ceramics *Eur. Phys. J.* B **89** 1–5
[29] Adamczyk-Habrajska M, Goryczka T, Szalbot D, Dzik J, Rerak M and Bochenek D 2020 Influence of lanthanum dopant on the structure and electric properties of $BaBi_2Nb_2O_9$ ceramics *Arch. Metall. Mater.* **65** 207–14
[30] Afqir M, Tachafine A and Fasquelle D 2018 Preparation and dielectric properties of $BaBi_{1.8}Ln_{0.2}Nb_2O_9$ (Ln = Ce, Gd) ceramics *Mater. Sci.-Poland* **36** 46–50
[31] Adak M K, Mukherjee A, Chowdhury A, Khatun J, Dhak P and Dhak D 2018 Effect of Cu^{2+} substitution on the relaxor behavior of $BaBi_2Nb_2O_9$ ferroelectric nanoceramics prepared by chemical route *Mater. Res. Express* **6** 025003
[32] Szalbot D, Adamczyk M, Wodecka-Duś B, Dzik J, Rerak M and Feliksik K 2018 Influence of calcium doping on microstructure, dielectric and electric properties of $BaBi_2Nb_2O_9$ ceramics *Process. Appl. Ceram.* **12** 171–9
[33] Adamczyk M, Ujma Z and Pawełczyk M 2006 Dielectric properties of $BaBi_2Nb_2O_9$ ceramics *J. Mater. Sci.* **41** 5317–22
[34] Ujma Z, Adamczyk M and Hańderek J 1998 Relaxor properties of $(Pb_{0.75}Ba_{0.25})(Zr_{0.7}0Ti_{0.30})O_3$ ceramics *J. Eur. Ceram. Soc.* **18** 2201–7
[35] Bräunlich P 1979 *Thermally Stimulated Relaxation in Solids* (Berlin: Springer) **vol 37**
[36] Pradhan D K, Choudhary R N P, Rinaldi C and Katiyar R S 2009 Effect of Mn substitution on electrical and magnetic properties of $Bi_{0.9}La_{0.1}FeO_3$ *J. Appl. Phys.* **106** 024102
[37] Skanavi G I and Matveeva E N 1957 New nonpiezoelectric dielectrics with very high dielectric permeability and small conductivity *Sov. Phys. JETP* **3** 905–8
[38] Burn I and Neirman S 1982 Dielectric properties of donor-doped polycrystalline $SrTiO_3$ *J. Mater. Sci.* **17** 3510–24
[39] Kao K C 2004 *Dielectric Phenomena in Solids* (Amsterdam: Elsevier)
[40] Lanfredi S and Rodrigues A C M 1999 Impedance spectroscopy study of the electrical conductivity and dielectric constant of polycrystalline $LiNbO_3$ *J. Appl. Phys.* **86** 2215–9
[41] Sinclair D C and West A R 1989 Impedance and modulus spectroscopy of semiconducting $BaTiO_3$ showing positive temperature coefficient of resistance *J. Appl. Phys.* **66** 3850–6
[42] Irvine J T, Sinclair D C and West A R 1990 Electroceramics: characterization by impedance spectroscopy *Adv. Mater.* **2** 132–8
[43] Hodge I M, Ingram M D and West A R 1976 Impedance and modulus spectroscopy of polycrystalline solid electrolytes *J. Electroanal. Chem. Interf. Electrochem.* **74** 125–43
[44] Rahman M A and Hossain A A 2014 Relaxation mechanism of $(x)Mn_{0.45}Ni_{0.05}Zn_{0.50}Fe_2O_4 + (1-x)BaZr_{0.52}Ti_{0.48}O_3$ multiferroic materials *Phys. Scr.* **89** 115811
[45] Handerek J, Ujma Z, Carabatos-Nédelec C, Kugel G E, Dmytrow D and El-Harrad I 1993 Dielectric, pyroelectric, and thermally stimulated depolarization current investigations on lead–lanthanum zirconate–titanate-*x*/95/5 ceramics with La content $x = 0.5\%$–4% *J. Appl. Phys.* **73** 367–73
[46] Bekhtin M A, Bush A A, Kamentsev K E and Segalla A G 2016 Preparation and dielectric and piezoelectric properties of Bi_3TiNbO_9, $Bi_2CaNb_2O_9$, and $Bi_{2.5}Na_{0.5}Nb_2O_9$ ceramics doped with various elements *Inorg. Mater.* **52** 510–6

[47] Zubkov S V 2020 Crystal structure and dielectric properties of layered perovskite-like solid solutions $Bi_{3-x}Gd_xTiNbO_9$ ($x = 0, 0.1, 0.2, 0.3$) with high Curie temperature *J. Adv. Dielectr.* **10** 2060002

[48] Zubkov S V and Vlasenko V G 2017 Crystal structure and dielectric properties of layered perovskite-like solid solutions $Bi_{3-x}Y_xTiNbO_9$ ($x = 0, 0.1, 0.2, 0.3$) with high Curie temperature *Phys. Solid State* **59** 2325–30

[49] Zhou Z, Dong X, Yan H, Chen H and Mao C 2006 Doping effects on the electrical conductivity of bismuth layered Bi_3TiNbO_9-based ceramics *J. Appl. Phys.* **100** 044112

[50] Zhang Y, Huang P, Zhu L, Du J, Bai W, Lin M, Zheng P, Zheng L and Zhang Y 2020 Doping level effects in Nb self-doped Bi_3TiNbO_9 high-temperature piezoceramics with improved electrical properties *Int. J. Appl. Ceram. Technol.* **17** 2407–15

[51] Peng Z, Yan D, Chen Q, Xin D, Liu D, Xiao D and Zhu J 2014 Crystal structure, dielectric and piezoelectric properties of Ta/W co-doped Bi_3TiNbO_9 Aurivillius phase ceramics *Curr. Appl Phys.* **14** 1861–6

[52] Wang Q, Wang C M, Wang J F and Zhang S 2016 High performance Aurivillius-type bismuth titanate niobate (Bi_3TiNbO_9) piezoelectric ceramics for high temperature applications *Ceram. Int.* **42** 6993–7000

[53] Basheer M A, Prasad G, Kumar G S and Prasad N V 2019 Dielectric studies on Sm-modified two-layered BLSF ceramics *Bull. Mater. Sci.* **42** 1–11

[54] Yan H, Zhang H, Zhang Z, Ubic R and Reece M J 2006 B-site donor and acceptor doped Aurivillius phase Bi_3TiNbO_9 ceramics *J. Eur. Ceram. Soc.* **26** 2785–92

[55] Subbarao E C 1962 A family of ferroelectric bismuth compounds *J. Phys. Chem. Solids* **23** 665–76

[56] Wang C M, Zhang S, Wang J F, Zhao M L and Wang C L 2009 Electromechanical properties of calcium bismuth niobate ($CaBi_2Nb_2O_9$) ceramics at elevated temperature *Mater. Chem. Phys.* **118** 21–4

[57] Mahajan S, Thakur O P, Prakash C and Sreenivas K 2011 Effect of Zr on dielectric, ferroelectric and impedance properties of $BaTiO_3$ ceramic *Bull. Mater. Sci.* **34** 1483–9

[58] Subohi O, Bowen C R, Malik M M and Kurchania R 2016 Dielectric spectroscopy and ferroelectric properties of magnesium modified bismuth titanate ceramics *J. Alloys Compd.* **688** 27–36

[59] Ismailzade I H, Yakupov R G and Melik-Shanazarova T A 1971 The magnetoelectric effect in ferroelectric–antiferromagnetic $Bi_5Bi_4Ti_3Fe_5O_{27}$ *Phys. Status Solidi* a **6** K85–7

[60] Kim S K, Miyayama M and Yanagida H 1996 Electrical anisotropy and a plausible explanation for dielectric anomaly of $Bi_4Ti_3O_{12}$ single crystal *Mater. Res. Bull.* **31** 121–31

[61] Hervoches C H, Snedden A, Riggs R, Kilcoyne S H, Manuel P and Lightfoot P 2002 Structural behavior of the four-layer Aurivillius-phase ferroelectrics $SrBi_4Ti_4O_{15}$ and $Bi_5Ti_3FeO_{15}$ *J. Solid State Chem.* **164** 280–91

[62] Singh R S, Bhimasankaram T, Kumar G S and Suryanarayana S V 1994 Dielectric and magnetoelectric properties of $Bi_5Ti_3FeO_{15}$ *Solid State Commun.* **91** 567–9

[63] Kubel F and Schmid H 1992 X-ray room temperature structure from single crystal data, powder diffraction measurements and optical studies of the Aurivillius phase $Bi_5Ti_3FeO_{15}$ *Ferroelectrics* **129** 101–12

[64] Mohapatra A, Das P R and Choudhary R N P 2014 Structural and electrical properties of $Bi_5Ti_3FeO_{15}$ ceramics *J. Mater. Sci., Mater. Electron.* **25** 1348–53

[65] Jartych E, Pikula T, Mazurek M, Franus W, Lisinska-Czekaj A, Czekaj D, Oleszak D, Surowiec Z, Aksenczuk A and Calka A 2016 Structure and magnetic properties of $Bi_5Ti_3FeO_{15}$ ceramics prepared by sintering, mechanical activation and Edamm process. A comparative study *Arch. Metal. Mater.* **61** 869–74
[66] Bai W, Gao Y Q, Zhu J Y, Meng X J, Lin T, Yang J, Zhu Z Q and Chu J H 2011 Electrical, magnetic, and optical properties in multiferroic $Bi_5Ti_3FeO_{15}$ thin films prepared by a chemical solution deposition route *J. Appl. Phys.* **109** 064901
[67] Silva P H T, Silva M A S, da Silva R B, Correa M A, Bohn F, de Menezes A S, Ferreira W C, Ayala A P, Sombra A S B and Fechine P B A 2020 Effects of the Bi^{3+} substitution on the structural, vibrational, and magnetic properties of bismuth layer-structured ferroelectrics *Appl. Phys.* A **126** 1–12
[68] Chen X, Xiao J, Xue Y, Zeng X, Yang F and Su P 2014 Room temperature multiferroic properties of Ni-doped Aurivillius phase $Bi_5Ti_3FeO_{15}$ *Ceram. Int.* **40** 2635–9
[69] Nakashima S, Nakamura Y and Okuyama M 2007 Characterization of epitaxial bi-layer-structured $Bi_5Ti_3FeO_{15}$ ($m = 4$) and $Bi_4Ti_3O_{12}$–$Bi_5Ti_3FeO_{15}$ ($m = 3$–4) natural-superlattice-structured multiferroic thin films prepared by pulsed laser deposition *MRS Online Proc. Lib.* **1034** 19–24
[70] Bai Y, Chen J and Zhao S 2016 Magnetoelectric fatigue of Ho-doped $Bi_5Ti_3FeO_{15}$ films under the action of bipolar electrical cycling *RSC Adv.* **6** 41385–91
[71] Koval V, Shi Y, Skorvanek I, Viola G, Bures R, Saksl K, Roupcova P, Zhang M, Jia C and Yan H 2020 Cobalt-induced structural modulation in multiferroic Aurivillius-phase oxides *J. Mater. Chem.* C **8** 8466–83
[72] Palizdar M *et al* 2012 Crystallographic and magnetic identification of secondary phase in orientated $Bi_5Fe_{0.5}Co_{0.5}Ti_3O_{15}$ ceramics *J. Appl. Phys.* **112** 073919
[73] Ferrer P, Iglesias J E and Castro A 2004 Synthesis of the Aurivillius phase $SrBi_4Ti_4O_{15}$ by a mechanochemical activation route *Chem. Mater.* **16** 1323–9
[74] Xu Z, Chu R, Hao J, Yao Z and Li H 2017 Hydrothermal preparation and electrical properties of Aurivillius phase $SrBi_4Ti_4O_{15}$ ceramic *Ferroelectrics* **516** 148–55
[75] Kalaiselvi B 2010 Synthesis and characterization of $SrBi_4Ti_4O_{15}$ by wet chemical method *Int. J. Pure Appl. Phys.* **6** 207–12
[76] Mamatha B, Suresh M B, James A R, Vithal M and Sarah P 2011 Synthesis and electrical properties of $SrBi_4Ti_4O_{15}$ piezoelectric ceramics *Phys. Scr.* **84** 055704
[77] Shrivastava S, Subohi O and Malik M M 2013 Ferroelectric properties of strontium bismuth titanate ($SrBi_4Ti_4O_{15}$) synthesized using solution combustion technique *Nano Hybrids* **3** 67–79
[78] Badapanda T, Harichandan R K, Mishra A and Anwar S 2013 Relaxor ferroelectric behavior of $BaBi_4Ti_4O_{15}$ Aurivillius ceramic *J. Adv. Dielectr.* **3** 1350013
[79] Bobić J D, Petrović M V, Ilić N I, Palaimiene E, Grigalaitis R, Paiva-Santos C O, Cilence M and Stojanović B D 2015 Lead-free $BaBi_4Ti_4O_{15}$ ceramics: effect of synthesis methods on phase formation and electrical properties *Ceram. Int.* **41** 309–16
[80] Chakrabarti A, Bera J and Sinha T P 2009 Dielectric properties of $BaBi_4Ti_4O_{15}$ ceramics produced by cost-effective chemical method *Physica* B **404** 1498–502
[81] Qi W, Wang Y, Wu J, Hu Z, Jia C and Zhang H 2019 Relaxor ferroelectric and photocatalytic properties of $BaBi_4Ti_4O_{15}$ *Adv. Appl. Ceram.* **118** 418–24
[82] Coondoo I, Jha A K and Agarwal S K 2007 Enhancement of dielectric characteristics in donor doped Aurivillius $SrBi_2Ta_2O_9$ ferroelectric ceramics *J. Eur. Ceram. Soc.* **27** 253–60

[83] Kopp V S, Kaganer V M, Schwarzkopf J, Waidick F, Remmele T, Kwasniewski A and Schmidbauer M 2012 X-ray diffraction from nonperiodic layered structures with correlations: analytical calculation and experiment on mixed Aurivillius films *Acta Crystallogr.* A **68** 148–55

[84] Wang C M, Wang J F, Zhang S and Shrout T R 2009 Electromechanical properties of A-site (LiCe)-modified sodium bismuth titanate ($Na_{0.5}Bi_{4.5}Ti_4O_{15}$) piezoelectric ceramics at elevated temperature *J. Appl. Phys.* **105** 094110

[85] Wang C M, Wang J F and Gai Z G 2007 Enhancement of dielectric and piezoelectric properties of $M_{0.5}Bi_{4.5}Ti_4O_{15}$ (M = Na, K, Li) ceramics by Ce doping *Scr. Mater.* **57** 789–92

[86] Subbarao E C 1962 Crystal chemistry of mixed bismuth oxides with layer-type structure *J. Am. Ceram. Soc.* **45** 166–9

[87] Jannet D B, El Maaoui M and Mercurio J P 2003 Ferroelectric versus relaxor behaviour in $Na_{0.5}Bi_{4.5}Ti_4O_{15}$–$BaBi_4Ti_4O_{15}$ solid solutions *J. Electroceram.* **11** 101–6

[88] Supriya S, Kalainathan S and Swaroop S 2010 Synthesis and characterization of $Na_{0.5}Bi_{4.5}Ti_4O_{15}$ powders by stearic acid gel method *Arch. Appl. Sci. Res.* **2** 386–91

[89] Chen Y *et al* 2018 Enhanced dielectric and piezoelectric properties in $Na_{0.5}Bi_{4.5}Ti_4O_{15}$ ceramics with Pr-doping *Ceram. Int.* **44** 18264–70

[90] Kuai W J, Zhao M L, Wang C L, Wang L H, Wang J, Gai Z G, Zhang J L and Yin N 2013 Effect of composition on the structure and piezoelectricity of $Na_{0.5}Bi_{4.5}Ti_4O_{15}$-based flexoelectric-type polar ceramics *J. Alloys Compd.* **559** 76–80

[91] Raghavan C M, Kim J W, Song T K and Kim S S 2015 Microstructural and ferroelectric properties of rare earth (Ce, Pr, and Tb)-doped $Na_{0.5}Bi_{4.5}Ti_4O_{15}$ thin films *Appl. Surf. Sci.* **355** 1007–12

[92] Gao D, Kwok K W and Lin D 2011 Microstructure, piezoelectric and ferroelectric properties of Mn-added $Na_{0.5}Bi_{4.5}Ti_4O_{15}$ ceramics *Curr. Appl Phys.* **11** S124–7

[93] Lazarević Z, Romčević N and Bobić J 2009 Study of ferroelectric $BaBi_4Ti_4O_{15}$ obtained via mechanochemical synthesis *Optoelectron. Adv. Mater.-Rapid Commun.* **3** 700–3

[94] Kalaiselvi B 2010 Synthesis and characterization of $SrBi_4Ti_4O_{15}$ by wet chemical method *Int. J. Pure Appl. Phys.* **6** 207–12

[95] Gupta A, Reddy V R, Kurchania R and Subohi O 2019 Synthesis of $Na_{0.5}Bi_{4.5}Ti_4O_{15}$ by co-precipitation route and optimization of calcination and sintering temperatures *Appl. Phys.* A **125** 1–6

Chapter 4

Mixed layered Aurivillius phases

The ferroelectric properties for even ($2m$) and odd ($2m + 1$) layered Aurivillius compounds are different to each other. In odd layered Aurivillius oxides more than one phase transition temperature exists. The higher temperature corresponds to the ferroelectric-to-paraelectric transition, whereas the lower temperature corresponds to the ferroelectric-to-ferroelectric transition. Also, two axes of polarization exist, and a small degree of spontaneous polarization along the c-axis can be observed in addition to the major polarization along the a-axis. In even layered compounds the spontaneous polarization P_s is only along the a-axis. The polarization along the c-axis is cancelled due to mirror symmetry in these even layered oxides. Some of these layered oxides are relaxor ferroelectrics if the A-site is occupied by Ba/lanthanide ions and some are multiferroics if the B-site is occupied by Fe ions [1].

4.1 Structure of intergrowths

The mixed layered Aurivillius phase compounds are composed of the regular intergrowth of one half of the unit cell of an m member structure and one half of the unit cell of an $m + 1$ member structure. The constituent structures ($2m$ and $2m + 1$) arrange themselves alternately along the common c-axis. The m and $m + 1$ layers are displaced with respect to the Bi_2O_2 layers towards the plane (110) and separated by $a_o/\sqrt{2}$, where a_o is a parameter of a regular perovskite cell. The compounds with this type of structure are called mixed Aurivillius phases and they are described by the general formula $A_{2m-2}Bi_4B_{2m}O_{6m+6}$ where the A-site is occupied by mono-, di-, or tri-valent ions, such as Na^+, K^+, Ca^{2+}, Sr^{2+}, Ba^{2+}, Pb^{2+}, etc, or a combination of them, and the B-site is occupied by small, higher valent tetra-, penta-, or hexa-valent cations, such as Ti^{4+}, Nb^{5+}, and Ta^{5+}.

The typical intergrowth structures are Bi_2WO_6–Bi_3TiNbO_9 consisting of single-layered Bi_2WO_6 and double-layered Bi_3TiNbO_9 making the average layer number $m = 1.5$, similarly Bi_3TiNbO_9–$Bi4Ti_3O_{12}$ (BTN-BIT) with double-layered

Bi_3TiNbO_9 and three-layered $Bi4Ti_3O_{12}$ with $m = 2.5$ and $Bi_4Ti_3O_{12}$–$Bi_5Ti_3FeO_{15}$ ($m = 3.5$ layers) oxides.

The intergrowth Aurivillius phases have attracted much attention due to their interesting crystal structure and enhanced ferroelectric properties compared to their individual compounds. The mismatch between the two parent blocks may be considered as the reason for the improvement in the ferroelectric/piezoelectric properties of these intergrowth Aurivillius phases. Figure 4.1 shows a schematic representation of the crystal structure of Bi_3TiNbO_9–$Bi4Ti_3O_{12}$.

4.2 Ordering and disordering in intergrowths

Ordering and disordering in intergrowth structures is a phenomenon for separate study and analysis, as it becomes the cause for materials exhibiting novel properties. Locally disordered intergrowth structures (DOISs) are found even in systems forming ordered intergrowth structures (OISs) with long periodicities [3–5]. In addition, there is always a small degree of ordering even in DOISs [6–10]. Also, it is not necessary that the OISs are equilibrium structures and the nature of the DOISs is still to be explored. In fact, the DOISs can be regarded as special solid solutions of two or more different components (111). Therefore, the nature of the local ordering turns out to be key to understanding the DOISs. Kohn and Eckart [11, 12] considered that the local ordering could be recognized as an independent structure if it repeats more than five times. In contrast, Rao [13] postulated that the structure expected from the periodicities observed in HRTEM lattice images should be consistent with the unit cell dimensions from diffraction measurement. In the Aurivillius family, the disorder is associated with stacking faults that are caused by the continuous sequences of one parent structure or constituent layer.

Gao *et al* [14] observed local BTNO and BIT stacking faults in the ordered –23- stacking sequence of the Bi_3TiNbO_9–$Bi_4Ti_3O_{12}$ (BTN_1BIT) structure using HRTEM. They observed that stacking faults may also signify new stacking orders and the local ordering of –223- and –233-stacking with 2–3 repeats. This may be regarded as a sign of new recurrent Aurivillius intergrowths. Based on a systematic analysis of the number of various growth faults and their tendencies with sintering temperatures, a structural reorganization model was formulated using a common

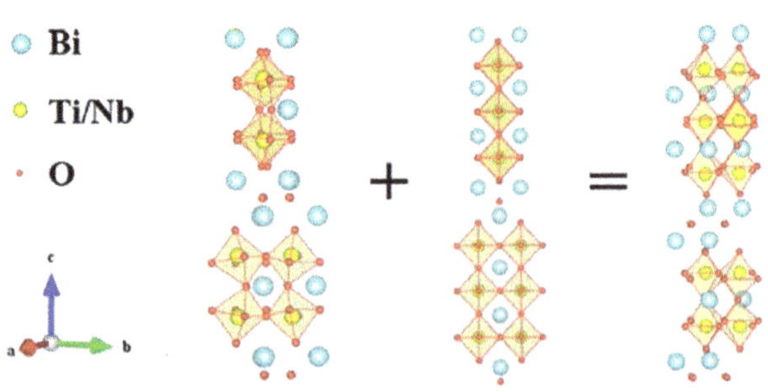

Figure 4.1. Prototype of the crystal structure of intergrowth Bi_3TiNbO_9–$Bi4Ti_3O_{12}$. Reproduced from [2] with permission of The Royal Society of Chemistry.

solution–precipitation process to understand the formation of the intergrowth structure. The favoured dissolution of the poorly bonded interleaved $[Bi_2O_2]^{2+}$ sheets allowed the Bi-rich melts to reorganize the BTN and BIT layered-units, which were re-ordered into the intergrowth during precipitation, together with all the types of growth faults. This model improves and re-defines the solid-state reaction to construct the intergrowth as an evolution process that is largely impervious to the compositional manipulation at the macroscopic level. Such a model and rationalization are based on the new findings of: (a) various types of BTN and BIT stacking faults in the intergrowth structure, some may even form new intergrowth locally; (b) the co-growth of intergrowth BTN on BIT and that of BIT on intergrowth BTN at low and high sintering temperatures; and (c) the correlation of anisotropic intergrowth with a Bi-rich intergranular phase.

Earlier studies by Shibuya *et al* [15] report the preparation of recurrent Aurivillius intergrowth film with the −223-stacking sequence using a pulsed laser deposition (PLD) method, but the PLD target they used was a BTN_1BIT rather than BTN_2BIT pellet, and the changes in the structure with the oxygen pressure and substrate temperature makes it difficult to link these recurrent −223-structures directly with the local −223-ordering in ceramics. In addition, Boullay *et al* [16] carried out associated research on $(Bi_3TiNbO_9)_2$–$(Bi_4Ti_3O_{12})$ (BTN_2BIT) and (Bi_3TiNbO_9)–$(Bi_4Ti_3O_{12})_2$ (BTN-BIT_2) systems. In their study, the structures of these two systems did not show the expected long-range ordered −223- and −233-stacking sequences, but the local orderings of these two sequences were observed.

As a typical case of recurrent intergrowth, the long-range-disordered BTN_2BIT (223) ceramics were reactive sintered and found to be composed of local −23-, −223-, −2223-, and −22-orderings. Figures 4.2(a) and (b) show high-resolution HAADF images of local −23- and −223-sequences, and figure 4.2(c) depicts the HAADF

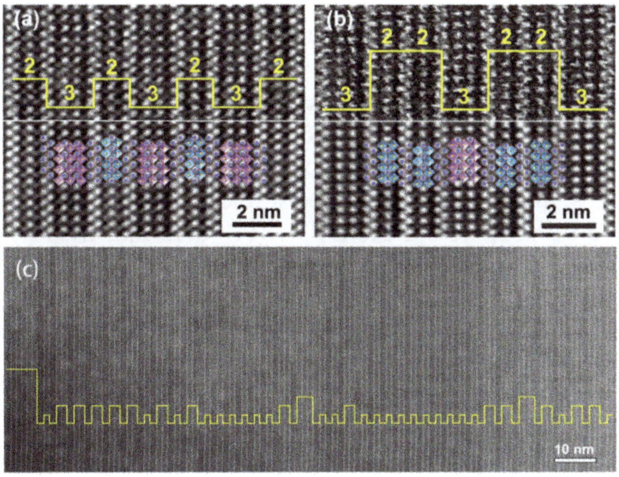

Figure 4.2. High-resolution HAADF images of local (a) −23- and (b) −223-sequences. (c) Medium-magnification HAADF image of a long-range-disordered structure in BTN_2BiT grains. Reproduced from [17]. Copyright (2014), with permission from Elsevier.

images of a long-range-disordered structure in BTN_2BIT_1 grains. To understand the intrinsic state of the system and to confirm its thermodynamic stability, Zhang *et al* [17] studied a long-range disordered BNT_2BIT system. They prepared a series of BNT_2BIT ceramics using the conventional mixed oxide route, with the powders being pre-sintered at 850 °C and then the pellets being reactive sintered at different temperatures from 950 °C to 1125 °C. They also grew single crystals of BNT_2BIT using a self-flux method. The process of phase evolution was studied using XRD patterns taken at different stages of sample preparation. It was observed that the pre-sintered powders contained only the parent phases BTN and BIT. The reactive sintering at higher temperatures induces the process of the intergrowth reaction among the parent phases. Based on a methodical examination of the statistical distribution of local ordering, the –223-structure was found to be the intrinsic state of this intergrowth system. The prepared BTN_2BIT single crystals were also found to possess –223-structure, which further confirms that the long-range-ordered 223-structure is the thermodynamic stable state of the BTN_2BIT system. Moreover, the nature of the local ordering was also discussed, and it was recommended that the repeating unit of $L = 4$ is the lower limit to uniquely determine an independent intergrowth structure, and the uncertain nature of the local –223-units with $L \leqslant 3$ is closely related to the vagueness in definition with associated orderings. It was also inferred that the kinetic process is the controlling factor to limit structural continuity and induce long-range-disordered intergrowth structures.

To overcome the limitation of the kinetic effect through multiple reactive sintering processes, Zhang *et al* [18] synthesized an abnormal intergrowth of a $BTN_1BiT_{2\text{–}233}$-structured ceramic using a multiple solid-state reactive sintering method. The microstructure, dielectric, piezoelectric, ferroelectric, and electrical properties of such compounds were studied methodically. A high-resolution STEM study showed that the long-range-ordered –233-sequences dominated the grain structure. The prepared sample was found to possess an orthorhombic crystal structure with space group $A2_1am$. The crystal structure was resolved by the software Expo2013 and MS programs from an artificial model to determine the lattice cell parameters as a = 5.458 Å, b = 5.415 Å, and c = 91.03 Å, respectively. Two dielectric anomalies attributed to the structure regularization of the BTN and BIT parts were observed at 668 °C and 760 °C. The d_{33} value (13.6 pC N^{-1}) of the sample was found to be higher than that of the solid-state reactive sintered BIT and BTN_1BiT_1 ceramics, and it was found to be stable up to the temperature 600 °C. The complete ferroelectric property of the BTN_1BIT_2 ceramic, with P_r = 13.4 μC cm^{-2} and E_c = 89.4 kV cm^{-1}, was also superior to the BTN_1BIT_1 ceramics. These results suggest that –233-structured BTN_1BiT_2 has good prospects both in FeRAM and high temperature piezoelectric devices. Therefore it can be stated that abnormal intergrowth structures with non-equal parent blocks will be potential candidates for improving the properties of Aurivillius oxides.

4.3 Non-successive layer intergrowths

As discussed, the natural intergrowth or superlattice Aurivillius phases consist of a combination of two Aurivillius phases with successive m numbers. It is interesting to note that the superlattice of the Aurivillius phases of non-successive *m* number show crystalline stacking and also ferroelectric properties. A study reported by Luo *et al* shows that the Bi_2WO_6–$Bi_4Ti_3O_{12}$ (m = 1–3) ferroelectric system does not develop into a 1–1 natural-superlattice structure but grows into a two-layered Aurivillius compound $Bi_3Ti_{1.5}W_{0.5}O_9$ [19]. They confirmed the formation of these two-layered compounds through the Rietveld refinement of the obtained XRD pattern. They further studied the mixture of two-layered $Bi_3Ti_{1.5}W_{0.5}O_9$ and three-layered $Bi_4Ti_3O_{12}$ in the ratio 1:3 and found that it is composed of an intergrowth with m = 2–3. The dielectric properties of all three phases were studied as a function of temperature and are shown in figure 4.3. It was observed that the Curie temperature of the 2–3 intergrowth is lower than that of the individual components $Bi_3Ti_{1.5}W_{0.5}O_9$ and $Bi_4Ti_3O_{12}$.

However, an intergrowth is formed with non-sequential m = 1, 3 with another member of one-layered Aurivillius phases, BMO, and three-layered BIT. Bi_2MoO_6–$Bi_4Ti_3O_{12}$ does grow into a natural superlattice with enhanced ferroelectric properties. XRD studies of Bi_2MoO_6–$Bi_4Ti_3O_{12}$ reported by Shibuya *et al* [20] clearly indicate that Bi_2MoO_6–$Bi_4Ti_3O_{12}$ does not form into a two-layered $Bi_3Ti_{1.5}Mo_{0.5}O_9$, but both the phases BMO or BIT appear separately. In addition, the crystal structures of BMO-BIT ceramics and thin films were found to be different. BMO-BIT thin films deposited on a $SrTiO_3$ single crystal substrate with pulsed laser deposition using the BMO-BIT ceramic as the target were single-phase with the lattice constant *c* found to be closer to that of the superlattice 1–1 structure with one unit of BMO and one unit of BIT as shown in figure 4.4. This superlattice of BMO-BIT with non-sequential *m* layers was first reported by Shibuya *et al.* The formation of the natural-superlattice 1–1 structure was confirmed by XRD reciprocal space mapping of BMO-BIT thin films. Thin films of BMO-BT intergrowth were also grown on Pt/TiO_2/SiO_2/Pt substrates at a temperature of 450 °C and post-annealed at 650 °C.

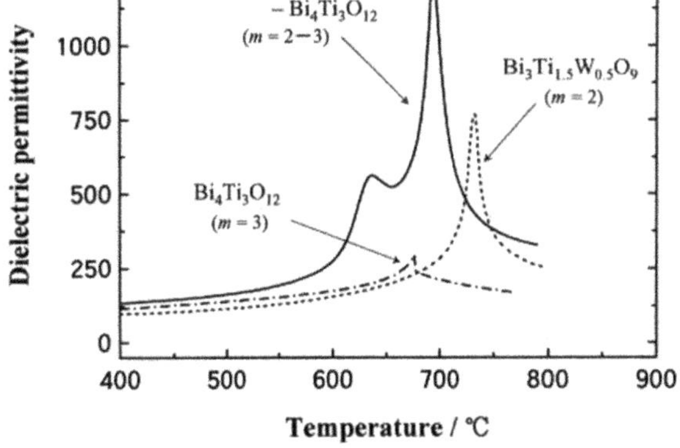

Figure 4.3. Variation of the dielectric constant with temperature for BIT, BTWO, and BTWO-BIT intergrowth. Reproduced from [19]. Copyright (2001) with permission from Elsevier.

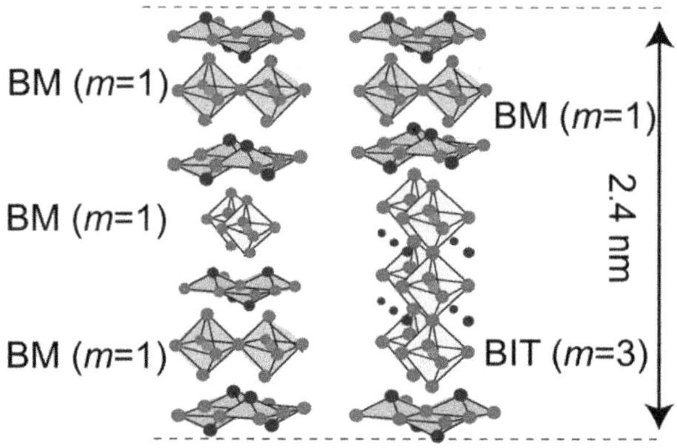

Figure 4.4. Prototype of the 1–3 BMO-BIT intergrowth structure. [20] 2006, reproduced by permission of the publisher (Taylor and Francis).

Enhanced ferroelectric properties with high remanent polarization and saturated *P–E* loops were obtained for post-annealed films.

4.4 Multiferroic intergrowths

The 2–3 and 3–4 intergrowth structure Bi_3TiNbO_9–$Bi_3LaTiNbFeO_{12}$ and $Bi_3LaTiNbFeO_{12}$–$Bi_5Ti_3FeO_{15}$ compounds were synthesized using a solid-state reaction method by Yang *et al* [2]. A superlattice-like distortion in these oxides was proposed, resulting from the combination of sublattices with different a and b lattice parameters, which was corroborated by XRD refinements and Raman spectra. Large variation in the c parameters for the samples with $m = 2.5$ and 3.5 was observed, which signifies the change in inter-layer length and the formation of intergrowth. This induced internal stress at the interface of the two constituent lattices thus affects its properties. The intergrowth was also validated by STEM-HAADF images as the 2–3 and 3–4 intergrowth structure. Shifts in the Raman peaks show the dependence of the Raman vibrations on the lattice parameters a and b, which confirms the existence of the superlattice structure. The *P–E* loops obtained suggest that the ferroelectric properties of the intergrowths are substantially enhanced compared to those of individual component oxides. The contribution of artificial polarization, such as that due to leakage current, was studied using pulsed polarization positive-up negative-down (PUND) measurements to validate that the superlattice-like structures improve the intrinsic ferroelectric polarization and coercive field of these oxides, specifically as compared to their component oxides Bi_3TiNbO_9, $Bi_3LaTiNbFeO_{12}$, and $Bi_5Ti_3FeO_{15}$. This enhancement of intrinsic polarization may be attributed to the mismatch of sublattices which induces interfacial tension in the intergrowth structure, resulting in larger BO_6 octahedral distortion and Jahn–Teller distortion. The Curie temperatures of the intergrowths were found to be lower than those of their constituent compounds due to the internal stress in the superlattice-like structure. Magnetic properties, such as magnetic hysteresis loops and the change in magnetization with temperature under field cooled and zero field cooled conditions, were studied. Bi_3TiNbO_9 was found to be diamagnetic while all the other compounds showed paramagnetism, which increases with Fe content. It was also found that the paramagnetism in these samples was dependent on the iron content, not on their intergrowth structure. Therefore, the magnetic properties remain unaffected by the structure.

The 3–4 layered intergrowth Aurivillius ceramics $Bi_4NdTi_3Fe_{1-x}Co_xO_{15}$–$Bi_3NdTi_2Fe_{1-x}Co_xO_{12-\delta}$ for $x = 0.1$, 0.3, 0.5, 0.7, 0.9, and 1 (BNTFC-x) with a 3–4 natural-superlattice structure were synthesized using conventional methods to examine the connection between the superlattice structure and properties [21]. By increasing the cobalt doping concentration, the four-layered parent phase steadily converts to a three-layered phase, in accordance with the structure evolution. The four-layered and three-layered phases can lock or modify each other via the lattice mismatch. Thus the observed properties cannot be explained by the simple collective effect of the two phases. As the cobalt doping level increases, the ferromagnetism appears in the ferroelectric material, and the remanent magnetization gradually increases and reaches a maximum value at $x = 0.3$, accompanied by the decrease of the remanent ferroelectric polarization. When the cobalt doping concentration is further increased, the remanent magnetization decreases along with the increase of ferroelectricity. Figure 4.5 shows the variation of remanent and saturation magnetization, remanent polarization, and coercive field with doping content x.

The magnetization of BNTFC-x is believed to be due to the contribution from the population of Fe^{3+}–O–Co^{3+} local structures as result of cobalt (Co) addition in the parent compound [22]. Moreover, the Ti/Fe/Co–O–Ti/Fe/Co angles might also change due to cobalt substitution [23] and this different structure growth [24] might be responsible for the observed magnetic properties. It is observed that the highest magnetization (M_r = 123 memu g^{-1}) obtained at $x = 0.3$ is about 31 times greater than that of $Bi_5Ti_3Fe_{0.5}Co_{0.5}O_{15}$ (3.9 memu g^{-1}) [22], and is comparable to that of $Bi_4NdTi_3Fe_{0.5}Co_{0.5}O_{15}$ (165 memu g^{-1}) [25] and $Bi_4NdTi_3Fe_{0.7}Ni_{0.3}O_{15}$ (194 memu g^{-1}) [26]. The variation of magnetic and ferroelectric properties can be explained well by the

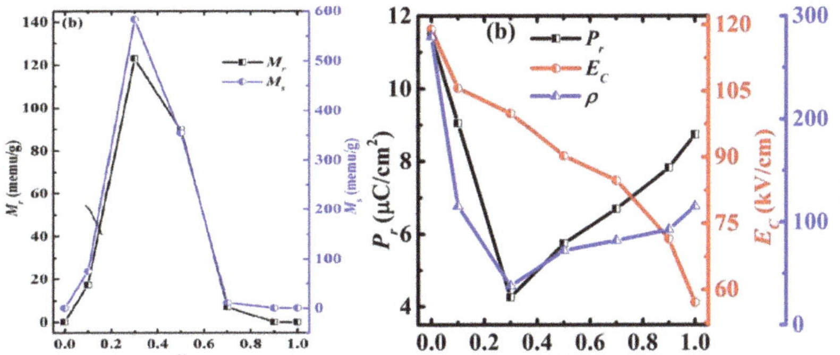

Figure 4.5. Variation of magnetization, remanent polarization coercive field, and resistivity of BNTFC-x intergrowth ceramics. [21] Copyright 2017. With permission of Springer.

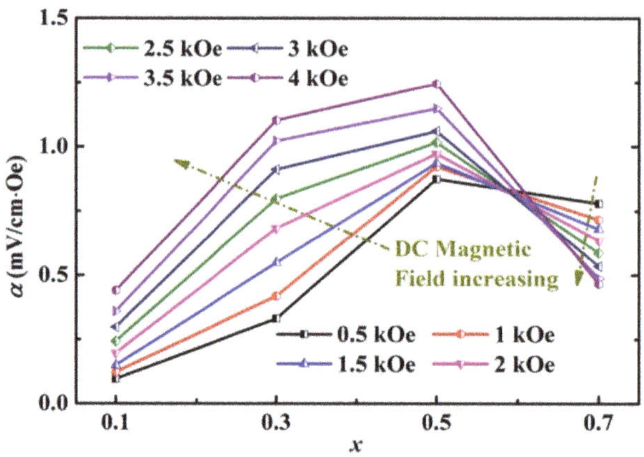

Figure 4.6. Variation of the magnetodielectric coefficient with the cobalt doping content. [21] Copyright 2017. With permission of Springer.

superlattice structure evolution. Furthermore, all the samples of BNTFC-x for $x = 0.1$–0.7 and possessing a ferromagnetic nature show a magnetoelectric (ME) effect under an ac magnetic field of 2.25 Oe with a frequency of 2 kHz at room temperature. The magnetoelectric coefficient was found to increase with an increase in applied dc magnetic field up to a doping concentration of $x = 0.5$. On further increasing the doping content a reversal in this trend was observed. The variation of magnetoelectric coefficient (α) with doping content (x) is shown in figure 4.6. This trend reversal may be attributed to the different responses of the magnetic and electric eigenmodes to the applied ac magnetic field frequency [27, 28].

4.5 Ferroelectric intergrowth ceramics

The synthesis of 2–3 intergrowth $BaBi_2Nb_2O_9$–$Bi_4Ti_3O_{12}$ ($BaBi_6Ti_3Nb_2O_{21}$) ceramic was first reported by Kikuchi *et al* [29]. They also reported the dielectric behaviour of this intergrowth as a function of temperature and found that all the 2–3 intergrowths possess two anomalies while in 3–4 intergrowths one peak is observed in their ε versus T curves measured at 1 MHz. However, no explanation was provided for the same. The frequency-dependent behaviour of the ε versus T curves for $BaBi_2Nb_2O_9$–$Bi_4Ti_3O_{12}$ intergrowth was first studied by Zhang *et al* [30]. They found that it is a relaxor ferroelectric and possesses the highest value of $T_m \sim 636$ °C at 100 kHz amongst all the other reported relaxor ferroelectrics (RFE). The dielectric properties of $BaBi_6Ti_3Nb_2O_{21}$ are considerably different from its two constituent oxides, $BaBi_2Nb_2O_9$ and $Bi_4Ti_3O_{12}$. It was observed that the dielectric peak at lower temperatures 200 °C–300 °C is frequency-independent, whereas the high temperature peak is frequency-dependent and shifts towards lower temperatures with a decrease in frequency. The relaxor behaviour could be confirmed in this compound by the dielectric anomaly at about 636 °C. The higher transition temperature corresponds to the Curie point and the transition at low temperature may be regarded as a phase transformation within orthorhombic symmetry due to a change in the space group. An obvious frequency dependence of the dielectric response was observed at room temperature, as is the case for a typical relaxor. The detectable ferroelectric domain switch and measurable value of the piezoelectric constant d_{33} clearly specified the ferroelectric nature of the $BaBi_6Ti_3Nb_2O_{21}$ ceramics. The weak piezoelectric response above 636 °C indicates the existence of spontaneous polarization of BBTN ceramics above T_m, which is consistent with it being an RFE.

The 3–4 intergrowth $Bi_4Ti_3O_{12}$–$SrBi_4Ti_4O_{15}$ (BIT-SBT15) ceramics were synthesized using a mixed oxide route by Noguchi *et al* [31]. The single-phase formation of BIT-SBT15 with space group $P2_1am$ having a regular intergrowth structure was confirmed by the Rietveld analysis performed on its XRD data. The lattice parameters of the intergrowth were determined and it was found that the lattice parameters $b = 0.5445$ nm and $c = 3.692$ nm of the intergrowth were an average of its constituents BIT and SBT15, while the lattice parameter $a = 0.5426$ nm was found to be lower than that of its constituents BIT (0.5447 nm) and SBT15 (0.545 nm). The ferroelectricity in Aurivillius oxides generally originates due to the displacement of A- and B-site ions along the *a*-axis [32], so the parameter *a* is closely related to the ferroelectricity. The well saturated *P–E* loop (shown in figure 4.7) supports its ferroelectric nature with a large remanent polarization ($2P_r$) of 30 μC cm^{-2}. The variation of dielectric constant with temperature shows a single peak depicting a first-order ferroelectric phase transition at a high Curie temperature of 610 °C. The T_c of intergrowth BIT-SBT15 lies between that of its constituents BIT (675 °C) and SBT15 (510 °C). The shorter lattice parameter *a* and shift in T_c of BIT-SBT15 could be the result of the strong ferroelectric interaction between two different kinds of perovskite blocks in the intergrowth structure.

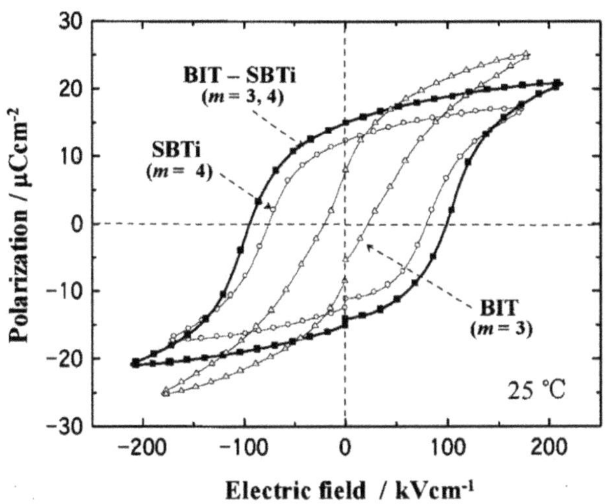

Figure 4.7. The *P–E* loop of BIT, SBT15, and BIT-SBT15 intergrowth ceramics measured at 25 °C. Reproduced from [31] with the permission of AIP Publishing.

The Rietveld analysis of the powder x-ray diffraction patterns showed that there are two kinds of Bi ions in Bi_2O_2 layers. One Bi ion in the Bi_2O_2 layers was displaced to the other Bi ion along the polarization direction by about 2% of parameter a, which is responsible for the larger $2P_r$. The lattice mismatch between the BIT blocks and SBT15 blocks in the intergrowth oxides causes major structural distortion in the Bi_2O_2 layers. Additionally, two kinds of Bi ions in the Bi_2O_2 layers for BIT-SBT15 are in different crystallographic surroundings—one bonds with the top oxygen of TiO_6 octahedra in the BIT blocks, while the other is connected with that in SBT15 blocks. These crystallographic factors lead to the displacement of Bi ions in the Bi_2O_2 layers.

The single-phase Bi_3TiNbO_9–$Bi_4Ti_3O_{12}$ (BTNO-BIT) ceramics belonging to 2–3 intergrowths were prepared using the conventional solid-state reaction route [33]. Its dielectric and ferroelectric properties were investigated. Two dielectric anomalies were observed in the temperature-dependent dielectric constant curves at 672 °C and 841 °C. The second anomaly is attributed to the ferroelectric-to-paraelectric transition and the first peak is ascribed to the restructuring of the BIT part of the intergrowth structure [34]. The ferroelectric property was evaluated using the P–E loop, with the obtained value of $2P_r = 20\ \mu C\ cm^{-2}$ and $2E_c = 184.6\ kV\ cm^{-1}$. The crystal structure analysis of the principal ferroelectric distortion, i.e. tilting and rotation of the octahedra, of the the intergrowth BTNO-BIT ferroelectric [35], and its constituent BTNO and BIT compounds [36, 37] was systematically studied. It was found that the component of the tilt angle (α) of the octahedra along the a–c plane (α_x) is responsible for the ferroelectricity of BTNO-BIT. The large P_s of the intergrowth compared to that of the BTNO and the easier domain switching under the electric field in comparison with that of the BIT results in the higher $2P_r$ for the BTNO-BIT ceramics than for the constituents.

The synthesis and properties of bismuth titanate ($Bi_4Ti_3O_{12}$) based intergrowths were studied by Machura *et al* [38]. Intergrowths $Bi_7Ti_4NbO_{21}$ with $m = 2.5$ and $SrBi_8Ti_7O_{27}$ with $m = 3.5$ comprising Bi_3TiNbO_9–$Bi_4Ti_3O_{12}$ and $Bi_4Ti_3O_{12}$–$SrBi_4Ti_4O_{15}$, respectively, were synthesized using the mixed oxide route. The phase formation and the layer structure were confirmed using x-ray diffraction analysis. The dependence of the dielectric constant on temperature showed two anomalies in 2–3 intergrowth $Bi_7Ti_4NbO_{21}$ at 657 °C and 847 °C, while only one anomaly was observed in 3–4 intergrowth $SrBi_8Ti_7O_{27}$ at 1007 °C. The piezoelectric coefficient (d_{31}) and electromechanical coupling coefficient (k_p) for both the intergrowths were evaluated as a function of temperature. It was observed that these coefficients remain constant up to temperatures of 350 °C and 200 °C for BTN and SBT, respectively. With a further increase in temperature d_{31} increases while k_p decreases.

Bismuth titanate ($Bi_4Ti_3O_{12}$)–strontium bismuth titanate ($SrBi_4Ti_4O_{15}$) ferroelectric intergrowth was synthesized using a modified oxalate route by Parida *et al* [39]. They studied the phase evolution of the intergrowth powders calcined at various temperatures from 300 °C to 900 °C using Rietveld refinement of x-ray diffraction data. It was observed that the low temperature calcined sample consisted of amorphous phases of Bi and Sr oxalates and hydroxides. At 300 °C the amorphous phases were converted into Bi–Sr oxide ($SrBi_2O_4$), Bi–C oxide, and Bi–Ti oxide. The Sr–Bi oxide phase was stable up to 500 °C and $Bi_4Ti_3O_{12}$ started forming only after 500 °C. No separate $SrBi_4Ti_4O_{15}$ phase was observed. It is believed to exist as a solid solution with BIT. The strongest peak at 30° was found to shift towards a higher angle with an increase in the calcination temperature from 600 °C to 900 °C. This shift in peak may be attributed to the formation of a greater and greater amount of BIT-SBT15 solid solution. Single-phase BIT-SBT15 intergrowth $SrBi_8Ti_7O_{27}$ was detected in samples calcined at 900 °C. The lattice parameters of the single-phase intergrowth were determined by performing Rietveld refinement of the XRD data using MAUD software, considering an orthorhombic crystal structure with space group *I2cm*. The lattice parameters of the intergrowth were found to be larger those of the individual components, indicating crystal mismatch and stresses induced in the intergrowth ceramics. Dielectric studies indicate that the T_c of the intergrowth is 610 °C, which lies between that of the individual constituents. The ferroelectric properties of the intergrowth are found to have improved compared to BIT and SBT15.

Intergrowth superlattice structural $Na_{0.5}Bi_{2.5}Nb_2O_9$–$Bi_4Ti_3O_{12}$ (NBN-BIT) ceramics were synthesized using the solid-state reaction method by Tian *et al* [40]. The formation of single-phase intergrowth was confirmed using x-ray diffraction. The lattice parameters, as calculated from XRD data, show that the value of the lattice parameter *c* of the intergrowth NBN-BIT is an average value of its constituents. The morphology of the obtained ceramics of the intergrowth was studied using SEM micrographs, as shown in figure 4.8.

It was observed that the grain size and the porosity in the ceramics corresponding to the intergrowth were lower than those of the individual components of the superlattice. Also, the micrographs clearly show the presence of two different phases in the intergrowth ceramics. In the temperature dependence of the dielectric permittivity, two anomalies were observed at 663 °C and 780 °C, respectively. The dielectric loss tanδ of the NBN-BIT ceramics was very low, in particular from

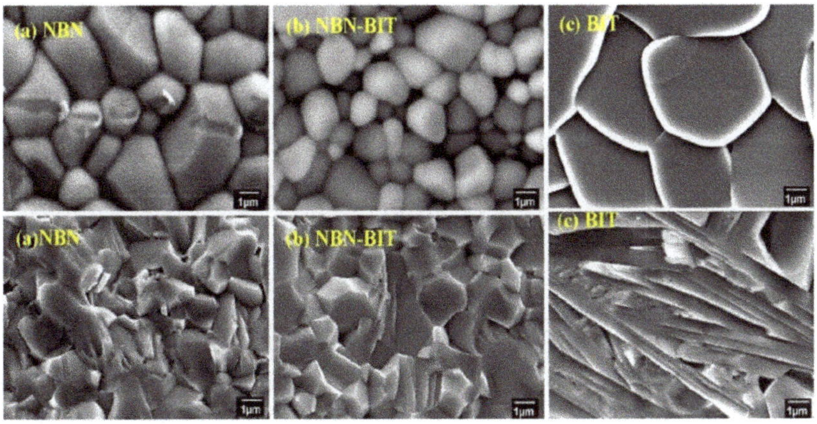

Figure 4.8. SEM micrographs of a natural surface and fractured surface. (a) NBN, (b) NBN-BIT, and (c) BIT ceramics. [40] Copyright 2014. With permission of Springer.

room temperature to 400 °C. The ferroelectric property was studied using the *P–E* hysteresis loop. A saturated loop was obtained for NBN-BIT intergrowth ceramics and the values of P_r and E_c were found to be 10.5 μC cm^{-2} and 18.5 kV cm^{-1}, respectively. It is observed that the P_r of the intergrowth is nearly twice as that of the NBN ceramics and larger than that of the BIT ceramics. Therefore the intrinsic ferroelectric property of NBN-BIT may be considered larger than that of NBN and BIT. The larger P_r and lower E_c of the NBN-BIT ceramics mean that the domain switching of NBN-BIT under an electric field is easier than that of the NBN and BIT. The results show that the ferroelectric properties of the NBN-BIT ceramics improved compared to pure NBN. Hence NBN-BIT intergrowth ceramics can be regarded as potential candidates for high temperature applications.

4.6 Ferroelectric intergrowth thin films

The $Bi_4Ti_3O_{12}$–$SrBi_4Ti_4O_{15}$ (BIT-SBT15) and $Bi_{4-x}La_xTi_3O_{12}$–$SrBi_4Ti_4O_{15}$ (BLT-SBT15) ferroelectric thin films were fabricated using a pulsed laser deposition (PLD) technique by Zhu *et al* [41]. The target pellets of BIT-SBT15 and BLT-SBT15 for PLD were prepared using the conventional solid-state route by ball milling the precursor mixture for 6 h followed by pre-sintering at 800 °C for 2 h and sintering the pressed pellets at 1200 °C for 2 h. The BIT-SBT15 and BLT-SBT15 thin films were then deposited on the Pt (111)/TiO_2/SiO_2/Si (100) substrate at 400 °C for 15 min using a krypton fluoride excimer laser with λ = 248 nm and 25 ns pulse duration. After deposition, the BIT-SBT15 films were annealed in a rapid thermal processor (RTP) at 650 °C and the BLT-SBT films were annealed at 700 °C for 5 min in an oxygen flow. The x-ray diffraction patterns of both the films indicate that they possess regular intergrowth structures and a single phase. SEM micrographs show that the films are around 250 nm thick and their surface is crack-free with a grain size of approximately 100–150 nm. A cross-sectional view of the film shows a clear demarcation between the the Pt electrode and the film. The ferroelectric behaviour of the films is studied using a *P–E* hysteresis loop measured at different voltages. A well saturated loop is obtained at a high voltage with remanent polarization P_r = 12 μC cm^{-2} and 13.5 μC cm^{-2} for the BIT-SBT15 and BLT-SBT15 thin films, respectively. Both of them show a well-defined hysteresis loop and good retention properties up to 10^5 s of retention time. The BIT-SBT15 shows fatigue behaviour, where a more than 40% decrease in remanent polarization was observed after switching cycles of 10^{10}, but a serious degradation of P_r was observed in the BIT-SBT15 films, making them unsuitable for FeRAM application. However, no obvious degradation of P_r was observed up to 10^{10} switching cycles in the BLT-SBT15 film at an applied field of above 250 kV cm^{-1}. The pinning of oxygen vacancies and field assisted domain unpinning leads to the strong field dependence of fatigue resistance. The chemical stability of the perovskite layer, oxygen vacancies, and anti-phase boundaries (APBs) existing in La doped BIT and SBT may be the main reason for the fatigue-free BLT-SBT15 thin film. The larger P_r, good retention, and fatigue-free properties demonstrate that BLT-SBT15 thin film is a promising candidate for FeRAM applications.

Natural-superlattice-structured/intergrowth $CaBi_2Nb_2O_9$–$Bi_4Ti_3O_{12}$ (CBNO-BIT) ferroelectric thin films were prepared successfully via a magnetron sputtering process by Xue *et al* [42]. The cylindrical disc shaped target BIT-CBN was calcined at 1120 °C for 10 h using a reactive sintering method and then annealed at 700 °C for 6 h for structure stabilization. The BIT-CBN thin films were deposited with a thickness of 200 nm on the MgO single crystal substrate by magnetron sputtering at a pressure of 2×10^{-4} Pa. In order to minimize the change in the film's composition from the stoichiometric $CaBi_6Nb_2Ti_3O_{21}$, due to surface decomposition (i.e. reduction of oxides) or a strong bulk diffusion of the sputtering target, a sputtering atmosphere with an Ar/O_2 flow ratio of 4:1 and a sputtering power of 70 W were chosen [43]. The phase formation and crystallographic orientation were studied using XRD and the film was found to be polycrystalline with a (200)/(020) texture, which can be ascribed to an improved lattice matching between (200)/(020) CBNO-BIT and the underlying (100) MgO substrate [43]. Atomic force microscope (AFM) micrographs of the CBNO-BIT film were analysed and the parameter root-mean-square (RMS) surface roughness of ~1.35 nm was determined. Such a smooth surface may be useful in the fabrication and integration of thin-film based microelectronic devices. The cross-sectional TEM image of the film shows a typical columnar grain structure with an average grain width of ~40 nm. Except for a preferentially thinned interface layer during the TEM sample preparation [44, 45], no other structural defects (voids, cracks, etc) with a size of ~10 nm or above could be observed in the TEM image. The lattice parameters calculated from the SAED patterns ($a = b = 0.5202$ nm, $c = 5.859$ nm) can be corroborated with the XRD data interpreted by using the intergrowth model. Second, the *c*-axis lattice parameter (5.859 nm) can be measured directly on both HRTEM images and is found to be the sum of the parent compounds CBN and BIT, indicating a natural-superlattice structure at the nanoscale. The XPS and EDS results confirmed that the film composition is close to chemical stoichiometry. The values of the binding energies obtained from the XPS data reveal that the elements Ca, Bi, Ti, and Nb exist in the Ca^{2+}, Bi^{3+}, Ti^{4+}, and Nb^{5+} valence states. With its microstructure successfully tailored at the nanoscale, the CBNO-BIT film exhibits good electrical properties, including a large dielectric constant ($\varepsilon_r \sim 390$), a high piezoelectric coefficient ($d_{33} \sim 90$ pm V^{-1}), as well as a high energy storage density ($W_E \sim 76$ J cm^{-3}). According to the equation governing the intrinsic piezoelectric d_{33} coefficient, i.e.

$$d_{33} = 2Q\varepsilon_o\varepsilon_r PE, \tag{4.1}$$

where Q, ε_o, ε_r, P, and E are the electrostrictive coefficient, permittivity in a vacuum, relative dielectric constant, polarization, and electric field, respectively. The increment in piezoelectric d_{33} coefficient can be typically ascribed to the enhanced dielectric constant (ε_r) and enhanced polarization (P). Figure 4.9 shows the piezoelectric strain loop (ε_3–E) and the computed piezoelectric d_{33} loop (d_{33}–E) of the CBNO-BIT film. The leakage current density was found to be lower than 2.0×10^{-6} A cm^{-2} up to an applied field of 290 kV cm^{-1}, indicating that the as-prepared film is

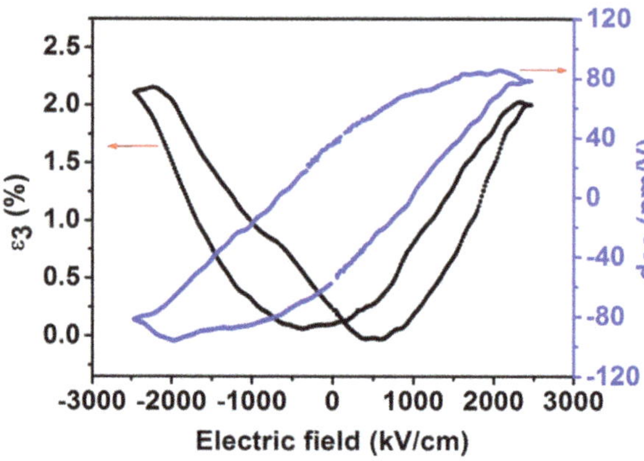

Figure 4.9. Piezoelectric strain loop (ε_3–E) and the computed piezoelectric d_{33} loop (d_{33}–E) of the CBNO-BIT film. Reproduced from [42], Copyright (2017) with permission from Elsevier.

insulating in nature. Two dielectric peaks were observed in the ε versus temperature curve with the first peak near 330 °C and the second peak near 480 °C. This is in accordance with the earlier reported results for the intergrowth ferroelectric ceramics [46–48]. However, the temperatures corresponding to the peaks are lower than those observed in the bulk ceramics by Duran *et al* [53].

$Bi_4Ti_3O_{12}$–$CaBi_4Ti_4O_{15}$ (BIT-CBT15) film was fabricated on Pt(111)/Ti/SiO_2/Si substrate using the sol–gel method. Yan *et al* [49] reported that the intergrowth structure was present in both the film and the corresponding powder sample from the XRD patterns. A good fatigue resistance as well as a strong charge-retaining ability were reported in the intergrowth BIT-CBT15 film. The remanent polarization (P_r) and coercive field (E_c) for BIT-CBT15 film was about 28 μC cm^{-2} and 150 kV cm^{-1} under an applied electric field of 540 kV cm^{-1}, respectively. The higher value of P_r may be attributed to the stress formed near the Bi_2O_2 layers between the two different layers of perovskite blocks as proposed by Noguchi *et al* [24, 31, 50] and it can also be ascribed to the large grain size of the obtained film. As calculated from the equation (4.1) the estimated value of P_r of purely (100)-oriented BIT-CBT15 film is expected to be greater than 50 μC cm^{-2} based on both the volume fraction of (100)-oriented grains and the piezoelectric properties. To elucidate this, a predominantly (100)-oriented BIT-CBT15 film was fabricated. As mentioned in the literature, the major polarization vector of Aurivillius oxides is close to the *a*-axis [51]. The XRD pattern of (100)-oriented BIT-CBT15 film confirms that the growth is along the *a*-axis and the strongest peak with a (100) orientation is obtained. The *P*–*E* loop of BIT-CBT15 (100)-oriented film shows that the remanent polarization is $P_r = 50$ μC cm^{-2} under an electric field of 1100 kV cm^{-1}. It implies that the theoretical prediction of the performance of (100)-oriented BIT-CBT15 films based on the ferroelectric and piezoelectric properties of the polycrystalline BIT-CBT15 film was reasonably valid. Moreover, the spontaneous polarization was found to be greater than 80 μC cm^{-2} under an applied electric field of 1100 kV cm^{-1}, which is around 60% higher than that of $Bi_4Ti_3O_{12}$ along its *a*-axis (~50 μC cm^{-2}) [52]. The *P*–*E* loop measured at different

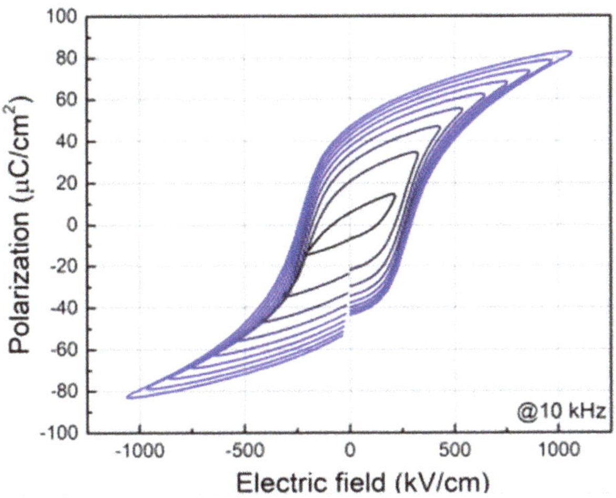

Figure 4.10. P–E loop measured at different applied electric fields for $Bi_4Ti_3O_{12}$–$CaBi_4Ti_4O_{15}$ film. Reproduced with permission from [49]. Copyright 2018 IOP.

applied electric fields (figure 4.10) shows that P_r and E_c increase rapidly at lower electric fields and then become saturated at higher electric fields (>200 kV cm^{-1}), indicating that the BIT-CBT15 film was well crystallized. The enhanced ferroelectric property may be ascribed to the formation of an intergrowth structure. The weak dependence of P_r on the frequencies ranging from 1 to 10 kHz demonstrates that the influence of the leakage current and nonlinear dielectric effects on the remanent polarization is very weak. The piezoelectric properties were studied using an atomic force microscope in piezoelectric mode. AFM micrographs show the formation of dense films with grain sizes ranging between 150–200 nm. Remarkably, the maximum piezo-response (d_{max}) in a single section line was about twice that of the average piezo-response (d_{ave}). This may be attributed to the (100)-oriented grains. As shown in figure 4.10, the hysteresis loops of the predominantly (100)-oriented BIT-CBT15 film show good rectangularity when the electric fields are higher than 500 kV cm^{-1}. This may also be ascribed to the well crystallized grains and high (100) orientation of the films.

4.7 The effect of doping in intergrowths

$Bi_{7-x}La_xTi_4NbO_{21}$ ($x = 0.00$, 0.25, 0.50, 0.75, 1.00, 1.25, 1.50, and 1.75) intergrowth ceramics were prepared using the traditional solid-state route with a pre-sintering temperature of 900 °C for 7 h and a sintering temperature of 1150 °C for 2 h by Yi *et al* [53]. The dielectric and ferroelectric properties of La doped intergrowth $Bi_7Ti_4NbO_{21}$ ceramics were investigated. The dielectric permittivity at a particular temperature is stable in a wide range of frequencies and changes with La^{3+} doping. Two dielectric anomalies were observed in the temperature-dependent dielectric constant curve of the undoped sample at around 670 °C and 840 °C. These two peaks are attributed to regularization of the $Bi_4Ti_3O_{12}$ phase and ferroelectric transition temperature, respectively. It was observed that these anomalies shift towards the low temperature side with an increase in La content. The shift in the peak positions is due to the replacement of the Bi^{3+} sites by La^{3+} as the 6s lone pair electrons of Bi influences the Curie temperature. In addition, these peaks tend to broaden and become diffusive for La content $x \geqslant 0.75$. The diffusiveness of the peaks for higher doping concentrations, i.e. $x \geqslant 0.75$ is due to the site disorder induced in $Bi_{7-x}La_xTi_4NbO_{21}$ ceramics. As Raman scattering is sensitive to the variations of local structures, the Raman shift with various La^{3+} contents gives valuable information about atomic or ionic substitution and polarization. When Bi^{3+} at the A-sites is substituted by lighter La^{3+} ions, the vibration modes 117 and 152 cm^{-1} shift towards higher frequencies. However, for higher doping content $x \geqslant 1.50$, there is no shift for these two modes, implying that within the detection limit of the instrument, no further substitution for Bi^{3+} at the A-sites takes place. Raman modes from 234 to 873 cm^{-1} are associated with the vibrations inside the perovskite-like layers, thus substituting La^{3+} for Bi^{3+} apparently affects these phonon modes. In the intergrowth $Bi_7Ti_4NbO_{21}$ structure with *I2cm* orthorhombic symmetry, there are four crystallographically inequivalent Bi sites [54]. Two of them (Bi1 and Bi2) are located at the A-sites in $(BiTiNbO_7)^{2-}$ and $(Bi_2Ti_3O_{10})^{2-}$ perovskite-like slabs, respectively. The rest (Bi3 and Bi4) are located at $(Bi_2O_2)^{2+}$ fluorite-like layers. When La^{3+} substitutes for Bi^{3+}, La^{3+} will preferentially enter the A-sites in the perovskite-like layers [55]. However, when a substantial amount of Bi^{3+} at the A-sites is substituted, the La^{3+} will inevitably enter the $(Bi_2O_2)^{2+}$ layers [56]. Substituting La^{3+} for Bi^{3+} at the A-sites affects the coordinate oxygen atoms located at the sharing-corner sites along the *ab*-plane and the internal apical sites of the octahedra simultaneously, whereas substituting La^{3+} for Bi^{3+} in the $(Bi_2O_2)^{2+}$ layers affects mainly the external apical atoms of the adjacent oxygen octahedral. Therefore, Raman scattering revealed that the Bi^{3+} in $(Bi_2O_2)^{2+}$ is replaced for $x \geqslant 1.00$ and there is no further Bi^{3+} substituted at the A-sites when $x \geqslant 1.50$.

Lanthanum modification in the intergrowth $Bi_4Ti_3O_{12}$–$SrBi_4Ti_4O_{15}$ ceramics was studied using Raman spectroscopy [57]. Ferroelectric ceramics of La doped $SrBi_4Ti_4O_{15}$ and La doped BIT-SBT15 ($SrBi_{7-y}La_yTi_8O_{27}$, BLT-SBLT15(y), $y = 0.00$, 0.25, 0.50, 0.75, 1.00, 1.25, and 1.50) were synthesized using a solid-state reaction technique and Raman spectrum measurements of the as-prepared ceramics were carried out at room temperature by Zhu *et al* [57]. The investigation on the Raman spectra suggests that La^{3+} ions prefer to occupy the A-site in the case of $x \leqslant 0.10$, and at higher content they tend to be incorporated into the Bi_2O_2 layers in the La doped $SrBi_4Ti_4O_{15}$. As for La doped $Bi_4Ti_3O_{12}$–$SrBi_4Ti_4O_{15}$, the critical content of the doping ions is 0.50. Beyond this doping level La enters the bismuth oxide layers. Based on the relationships of the Curie temperature or the lattice space for the BLT-SBLT15 intergrowth and its parent compounds, BLT and SBLT, the La distribution in La doped BIT-SBT15 was calculated. For BLT-SBLT15 ($y = 0.50$), the La content is 0.29~0.37 in the BLT unit and 0.13~0.21 in SBLT unit [58, 59]. The La content in the BLT unit is lower than the critical value in the La doped BIT (about 1.00) [19], while that in the SBLT unit almost reaches the critical value of La doped SBT15. In BIT-SBT15, the $(Bi_2O_2)^{2+}$ slabs are located between the pseudo-perovskite blocks of the BIT and SBT15 units, the $(Bi_2Ti_3O_{10})^{2-}$ and $(SrBi_2Ti_4O_{13})^{2-}$ blocks, along the *c*-axis [60]. As the La doping content is 0.50, the pseudo-perovskite blocks of the SBLT15 units reach the critical acceptability of La^{3+} ions. Thus the La^{3+} then start occupying the Bi sites in the $(Bi_2O_2)^{2+}$ slabs through the side of the $(SrBi_2Ti_4O_{13})^{2-}$ slabs. It is widely accepted that the $(Bi_2O_2)^{2+}$ layers have a very important influence on the properties of Aurivillius oxides. The $(Bi_2O_2)^{2+}$ layers behave as the insulating layers and are self-regulated to compensate for space charge due to their net electric charge [61, 62]. Substitution of La^{3+} for Bi^{3+} ions in the BLSFs will result in the reduction of the oxygen vacancies and the relaxation of the lattice distortion. As the La^{3+} ions are incorporated into the $(Bi_2O_2)^{2+}$ slabs, the effect of the insulated layers and the space charge compensation of the $(Bi_2O_2)^{2+}$ slabs become disturbed. The structure of the $(Bi_2O_2)^{2+}$ changes due to the incorporation of La^{3+} ions at the Bi sites, resulting in more substantial relaxation of the lattice distortion. Hence, in the La doped BLSTs, such as BIT, SBTi, and BIT-SBTi, the remnant polarization maximizes at the La content near which the La starts entering into the $(Bi_2O_2)^{2+}$ layers. Therefore, it can be concluded that the $(Bi_2O_2)^{2+}$ layers have a very important effect on the properties in Aurivillius oxides.

The effect of Er^{3+} doping in $Bi_4Ti_3O_{12}$–$SrBi_4Ti_4O_{15}$ intergrowth ceramics was investigated by Liu *et al* [63]. Polycrystalline samples of BIT-SBT-xEr^{3+} ($x = 0.00$, 0.05, 0.10, 0.15, and 0.20) ceramics were synthesized using the solid-state reaction method at a calcination temperature of 800 °C for 4 h and a sintering temperature of 1100 °C–1120 °C for 4 h. XRD patterns of Er doped BIT-SBT15 ceramics show the formation of single-phase compounds with no secondary phases present. They possess a orthorhombic structure with the *I2cm* space group. The maximum intensity peak in all the samples corresponds to the (118) plane in accordance with the strongest peak ($1\ 1\ 2m + 1$) where $m = 3.5$ in BLSF compounds [64]. SEM micrographs show the formation of plate-like grains in these Er^{3+} substituted BIT-SBT15 ceramics with a decrease in the average grain size due to Er^{3+} substitution, indicating that Er^{3+} inhibits grain growth. Raman spectroscopy was used to analyse the process of ion substitution. It was reported that the peaks corresponding to the frequencies 129 and 156 cm^{-1} shift towards the higher frequency side as the concentration of Er^{3+} increases. This shift in the peak positions may be attributed to the fact that as Er^{3+} ($A = 167.26$ a.m.u.) is lighter than Bi ($A = 208.98$ a.m.u.), the bond between Er–O will be stronger than Bi–O. Therefore, Er^{3+} is replacing Bi at the A-site in the perovskite block. The intensities of the modes at 346, 485, and 542 cm^{-1} increase with an increase in the Er^{3+} content. Two vibrational modes appear at 689 and 736 cm^{-1} and their intensity also increases as the Er^{3+} doping content increases. These changes indicate that there is an increase in the lattice distortion of Er^{3+}substituted BIT-SBT15 ceramics with an increase in Er^{3+} content [65]. The dielectric studies reveal that the dielectric constant and ferroelectric transition temperature T_c decreases with an increase in Er doping level. It is well known that the dielectric constant of a material depends on its ionic and electronic polarizability. The polarizability of Bi^{3+} ion is much larger due to the presence of the 6s lone pair electrons. As more Er^{3+} ions replace Bi^{3+} ions, a reduction in Bi-6s/O-2p hybridization occurs, which causes a substantial decrease in polarization [66, 67]. This caused a reduction in the dielectric constant of Er^{3+} substituted BIT-SBT15 ceramics. It was observed that the dielectric loss is much higher at high temperatures in BIT-SBT15 ceramics due to the creation of oxygen vacancies [68] as result of Bi volatilization at high temperatures. Er being more stable than Bi, its substitution at Bi sites reduces the oxygen vacancy formation thus reducing the dielectric loss in the ceramics. The conduction due to vacancy movement corresponding to the dc conductivity is one of the most prominent electrical properties of Aurivillius compounds. The activation energy for dc conductivity in the Er^{3+} doped BIT-SBT15 ceramics was evaluated through complex impedance spectroscopy. The Cole –Cole plots for various concentrations of Er doped ceramics are shown in figure 4.11. These plots were fitted by considering the equivalent circuit of a parallel combination of resistor, capacitor, and constant phase element (CPE, shown in the inset of figure 4.11). However, the role of diffused space charge on conductivity, such as mobile ions, dipoles, and hopping electrons, cannot be neglected for the actual crystal [69, 70]. The dc conductivity (σ_{dc}) can be evaluated by

$$\sigma = \frac{t}{R_{tot}A}, \tag{4.2}$$

where R_{tot} is obtained from the intercept of the curve on the real x-axis (Z'), and t and A represent the thickness and the area of the samples, respectively. The activation energy (E_a) for dc conductivity can be expressed by the Arrhenius relation. The oxygen vacancies V_O play a vital role in the electrical conductivity. Additionally, E_a increases with an increase in Er^{3+} doping content, signifying that a suitable Bi^{3+} ions replacement by Er^{3+} ions can decrease the concentration of V_O.

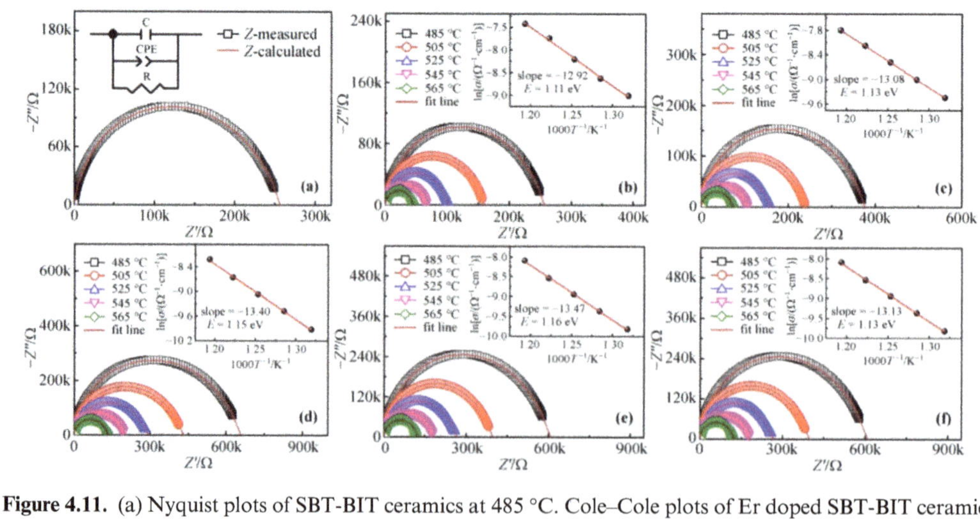

Figure 4.11. (a) Nyquist plots of SBT-BIT ceramics at 485 °C. Cole–Cole plots of Er doped SBT-BIT ceramics at different temperatures: (b) $x = 0.00$, (c) $x = 0.05$, (d) $x = 0.10$, (e) $x = 0.15$, and (f) $x = 0.20$. [63] Copyright 2019. With permission of Springer.

In lanthanum (La) doped $Bi_4Ti_3O_{12}$–$SrBi_4Ti_4O_{15}$ intergrowth ceramics the possibility of La^{3+} occupying the A-site in BIT and SBT15 is more due to their similar crystal structures. However, the La content in these two constituents may not be same due to the presence of Sr at the A-site in SBT15. The possibility of La occupying the Bi site is higher in both the constituents due to their similar ionic radii of Bi^{3+} (1.17 Å) and La^{3+} (1.16 Å). Therefore, it becomes imperative to study the distribution of La in BIT-SBT15 intergrowth to understand the crystal structure of La doped ceramics. Zhu *et al* [71] investigated the effect of La doping in BIT-SBT15 intergrowth ceramics on the crystal structure, and ferroelectric and dielectric properties. The $Bi_{4-x}La_xTi_3O_{12}$–$SrBi_{4-y}La_yTi_4O_{15}$ (BLT-SBLT($x + y$), $x + y$ = 0.00, 0.25, 0.50, 0.75, 1.00, 1.25, and 1.50) ceramic samples were prepared by the standard solid-state reaction method. The XRD patterns of BLT-SBLT ceramics show that all the samples are single-phase with intergrowth crystal structure. All the peaks for all values of $x + y$ coincide, implying that the crystal structure remains unaffected due to La addition. However, the maximum intensity peak around $2\theta = 30°$ shifts towards lower angles with an increase in La concentration up to $x + y = 1$. On further increasing the La content the shift in this peak shows the reverse trend. The ferroelectric properties of BIT-SBT15 were found to be improved remarkably by La doping. The value of remanent polarization $2P_r$ is found to be maximum for $x = 0.50$, an increase of approximately 60% compared to that of pure BIT-SBT15 intergrowth ceramic, whereas the increase in coercive field E_c is only 10%. The Curie temperature of BLT-SBLT($x + y$) intergrowth ceramics is found to decrease with increasing La doping. The T_c of pure BIT-SBT15 is 602 °C, which is the average T_c of BIT (675 °C) and SBT15 (520 °C). Thus, the T_c of La doped BIT-SBT15 for various concentrations of La will also be the average of that of BLT and SBLT given by

$$T_c(x + y) = \frac{T_{c1}(x) + T_{c2}(y)}{2}, \tag{4.3}$$

where $T_c(x + y)$, $T_{c1}(x)$, and $T_{c2}(y)$ are the Curie temperature of BLT-SBLT($x + y$) intergrowth, BLT(x), and SBLT(y), respectively, and $x + y$ is the total La content = 0.00, 0.25, 0.50, 0.75, 1.00, 1.25, and 1.50. The distribution La in BLT-SBLT intergrowth ceramics can be calculated from the fitted data of [72]

$$T_{c1} = 677 - 198x - 252x^2 \tag{4.4}$$

$$T_{c2} = 520 - 255y - 22y^2. \tag{4.5}$$

The value of La content in BLT-SBLT can be calculated from equations (4.3), (4.4), and (4.5). It was observed that the La content in BLT is higher than that in SBLT up to $x + y = 1.25$, whereas for $x = 1.5$ the La content in SBLT becomes higher than that in BLT. The difference in the distribution of La in the two constituents of the intergrowth might be due to the difference in their microstructure. As Sr^{2+} is also occupying the A-site in SBLT its presence changes the acceptability of other ions at that site. This explanation can also be corroborated with the change in d (118) with La doping level.

Cerium (Ce) doped $Bi_4Ti_3O_{12}$–$BaBi_4Ti_4O_{15}$ intergrowth ceramics were studied to understand the effect of Ce doping on its phase formation, crystal structure, morphology, dielectric properties, and piezoelectric properties by Xianwen *et al* [73]. $Ba_{1-x}Ce_xBi_8Ti_7O_{27}$ (x = 0, 0.05, 0.1, 0.15, and 0.2) ceramics were prepared using the conventional solid-state method. The XRD patterns of all the samples show that the samples possess single-phase crystal structure without the presence of any secondary phases. The lattice parameters of the Ce doped intergrowths were evaluated through Rietveld refinement. It was observed that the volume of the unit cell increases with an increase in Ce content, thus inducing orthorhombic distortion in the samples. This may be due to a difference in the ionic radii of the dopant and the host ion. SEM micrographs show that the ceramics possess plate-like structured grains with anisotropic shapes, which are typical of Aurivillius compounds. X-ray photoelectron spectroscopy of the as-prepared sample was carried out to understand its valence band position and chemical composition. It is noted from the XPS data that Ce ions exist in two valence states, Ce^{3+} and Ce^{4+}. Ce^{3+} may occupy Bi^{3+} sites and Ce^{4+} may occupy Ti^{4+} sites in the octahedral due to their similar ionic radii. It was also observed that the concentration of oxygen vacancies decreases for a lower doping concentration and then increases slightly with a further increase in Ce content. This may be due to the possibility of the occupation of the A-site by donor Ce^{4+} ions for a higher content of cerium. The possibility of Ce occupying both the A-site and B-site is also corroborated by Raman spectroscopy. The value of the dielectric constant decreases but T_c increases with an increase in the doping level of cerium. Ce doping in BIT-BBT15 intergrowth also induces a relaxor nature in the samples. Additionally, the Ce doped BIT-BBT15 intergrowth ceramics possess lower dielectric losses than the undoped ceramics, which may be due to the reduction in the amount of oxygen vacancies as observed through XPS spectra. The decrease in the pinning effect facilitates the movement of the domain wall and significantly enhances the piezoelectric properties due to the substantial reduction in oxygen vacancies. Among all the doped samples, the sample BBTI-BIT-0.1Ce showed the highest piezoelectric performance (20.6 pC N^{-1}) compared to that of the pure ceramics (6.7 pC N^{-1}).

B-site modification of $Bi_4Ti_3O_{12}$–$SrBi_4Ti_4O_{15}$ intergrowth ceramics by donor dopant tungsten (W^{6+}) was studied by Wang *et al* [74]. Among the various high valent dopants W^{6+} was chosen due to its similar ionic size as to that of Ti^{4+} and its superior doping effect in BIT. The basic crystal structure of BIT-SBT15 was not affected by a small amount of tungsten doping. $SrBi_{8-2x/3}Ti_{7-x}W_xO_{27}$ (BTW-SBTW (x), x = 0.00, 0.006, 0.03, 0.06, 0.09, and 0.12) ceramic samples were synthesized using the conventional solid-state reaction method. The XRD patterns show a single-phase intergrowth structure of the as-prepared series of compounds. Due to the similar ionic radius of W^{6+} (0.060 Å) and Ti^{4+} (0.0605 Å) the structural constancy of the samples is maintained. Well saturated *P–E* loops are obtained for series of doped and undoped BTW-SBTW intergrowth ceramics. The remanent polarization increases with an increase in W content up to a doping concentration of $x = 0.03$. The maximum value of $2P_r$ obtained for W doped BTW-SBTW ($x = 0.03$) ceramics was 43.2 $\mu C\ cm^{-2}$, around 2.5 times larger than that of undoped

BIT-SBT15. With a further increase in doping concentration, the remanent polarization decreases. The probable explanation for this behaviour is that the donor ion substitution not only reduces oxygen vacancies but also restricts its mobility, thus increasing the remanent polarization with an increase in dopant concentration up to $x = 0.03$. With a further increase in the dopant content bismuth vacancies are created according to the formula $SrBi_{8-2x/3}Ti_{7-x}W_xO_{27}$. These Bi vacancies are responsible for the deterioration of ferroelectric properties. Meanwhile, a very small change in coercive field is observed from 89.1 kV cm^{-1} to 94 kV cm^{-1}. This small change in E_c may be due to the very small lattice distortion induced in the ceramics by W substitution because of the similar ionic radii of W and Ti. Figure 4.12 shows the variation of P_r and E_c with tungsten doping content in $Bi_4Ti_3O_{12}$–$SrBi_4Ti_4O_{15}$ intergrowth ceramics. The dielectric properties of the W doped BIT-SBT15 intergrowth were studied as a function of temperature. It was observed that the dielectric constant increases but the Curie temperature decreases with an increase in tungsten doping content. The increase in dielectric constant with W content may also be attributed to fewer oxygen vacancies and lower vacancy mobility. The electrical properties were significantly enhanced by a small amount of W doping without compromising the thermal stability of the ceramic.

The effect of B-site Nb doping in $Bi_4Ti_3O_{12}$–$SrBi_4Ti_4O_{15}$ intergrowth ceramics prepared using a modified oxalate route on the phase formation microstructure, and dielectric, ferroelectric, and electrical properties was studied by Parida *et al* [64]. XRD phase analysis confirmed the formation of a single-phase intergrowth compounds. Nb doping does not affect the basic crystal structure of the intergrowth. SEM micrographs showed that the grain size of the ceramics decreases with Nb doping, indicating that niobium addition inhibits grain growth. The temperature dependence of the dielectric constant and loss was investigated in the temperature range 30 °C–800 °C and frequency range 1 kHz–1 MHz. With an increase in Nb doping, the T_c of the ferroelectrics decreases and the peak permittivity increases. Doping also introduces small relaxor behaviour in the ferroelectrics with a degree of diffuseness increasing with Nb content. The dc conductivity of the ceramics

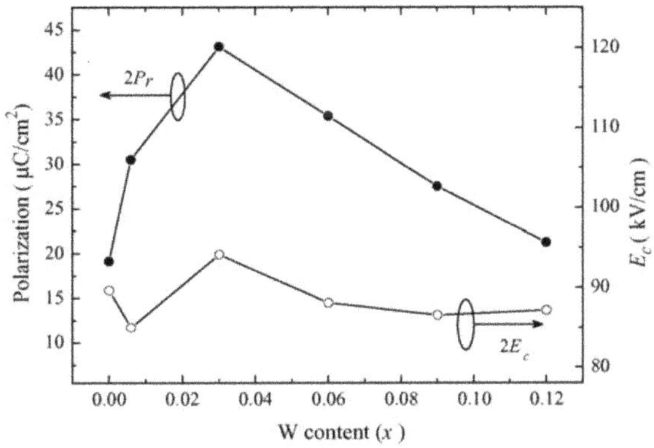

Figure 4.12. Variation of P_r and E_c with tungsten doping content in $Bi_4Ti_3O_{12}$–$SrBi_4Ti_4O_{15}$ intergrowth ceramics. Reproduced from [74], Copyright (2007) with permission from Elsevier.

decreases with doping. The remnant polarization (P_r) of the intergrowth ferroelectrics is increased with Nb doping. The oxygen vacancies play a vital role in tailoring the electrical properties of ceramics. The B-site donor doping reduces the oxygen vacancy concentration according to the following defect reaction:

$$Nb_2O_5 \leftrightarrow 2Nb_{Ti} + 5O + 2e \tag{4.6}$$

$$O_2(g) + V_O \leftrightarrow O_o. \tag{4.7}$$

The Nb doping produces two Nb_{Ti} holes and two electrons. These two electrons combine with one oxygen atom to balance an oxygen vacancy, resulting in a decrease in oxygen vacancy concentration. Accordingly, the dc activation energy calculated from the Cole–Cole plots increases with an increase in Nb content as a result of a reduction in oxygen vacancies.

Simultaneous A-site and B-site doping in $Bi_7Ti_4NbO_{21}$ intergrowth ceramics by heterovalent ions such as Sr (A-site) and W (B-site) was studied to understand the relationship between the crystal structure and ferroelectric properties by Ogawa *et al* [75]. The $Bi_{7-x}Sr_xTi_4Nb_{1-x}W_xO_{21}$ (BSTNW) for $x = 0$, 0.25, 0.5, and 0.75 ceramics were prepared using the solid-state reaction method at the calcination temperature of 800 °C and pressed into pellets then sintered at a temperature of 1100 °C for $x = 0$ and 0.25, and at 1170 °C for $x = 0.5$ and 0.75 for 2 h. The x-ray diffraction patterns of the as-prepared ceramics show the formation of single-phase compounds up to a doping content of $x = 0.5$ with no secondary phases present. The sample with $x = 0.75$ consists of secondary phases due to the decomposition of $Bi_7Ti_4NbO_{21}$ into $Bi_4Ti_3O_{12}$ and Bi_3TiNbO_9. The lattice parameters were calculated through Rietveld refinement for compositions from $x = 0$ to 0.5, while for $x = 0.75$ they were calculated by a least squares fit due to the existence of secondary phases in this ceramic. From the variation of the lattice parameters of the BSTNW ceramics with doping content, the solubility limit was determined to be $x = 0.5$. A significant change in the lattice parameter c is observed at $x = 0.5$, whereas little change is observed in lattice parameters a and b. Peaks tend to shift towards lower 2θ with an increase in doping level, indicating induced distortions in the samples due to differences in the ionic radii of Sr (1.4 Å) and Bi (1.3 Å). Furthermore, for the samples with doping content $x = 0$ to 0.5, it was observed that the Sr and W substitutions for Bi and Nb, respectively, lead to the enhancement of orthorhombic distortion in the oxygen octahedra. The ferroelectric properties were studied using *P–E* hysteresis loops for single-phase ceramics with $x = 0$–0.5. It was observed that remanent polarization increases with an increase in doping level. This improvement in the P_r values may be attributed to the induced structural distortion of oxygen octahedra by Sr and W substitutions for Bi and Nb, respectively. The maximum value of $P_r = 15.1\ \mu C\ cm^{-2}$ was obtained at $x = 0.5$. The variation of dielectric constant as function of temperature was also studied in these single-phase ceramics. Two dielectric anomalies were obtained. The dielectric constant increases while the T_c decreases with an increase in doping level. A decrease in T_c may be ascribed to an increase in the tolerance factor from 0.948 to 0.959 for doping content $i = 0$–0.5, which corresponds to an increased tilt in the octahedra. In addition, to understand the effect of the grain orientation of the ceramics on its ferroelectric properties, the oriented BSTNW ($x = 0.5$) ceramic was prepared using a hot-forging technique. The ferroelectric properties were measured on the sample in the direction perpendicular to the pressing direction and in the direction parallel to the pressing direction.

The XRD profiles of these hot forged BSTNW ceramics show strong reflections of the (0 0 *l*) plane, indicating growth along the *c*-axis in the pressing direction. The ferroelectric *P–E* loops measured (shown in figure 4.13) for these grain-oriented samples show that the measurements taken along the direction perpendicular to the pressing direction, i.e. along the *a–b* plane, show a much higher value of remanent polarization $P_r = 20.5\ \mu C\ cm^{-2}$ at 150 kV cm^{-1}.

Babu *et al* [76] prepared Sm doped mixed layered $Bi_4Ti_3O_{12}$–$SrBi_4Ti_4O_{15}$ intergrowth ferroelectrics. It was speculated that the substitution of Bi^{3+} (1.17 Å) with larger size cations Sm^{3+} (1.27 Å) can efficiently stabilize the oxygen ion in the lattice and leads to an improvement in the dielectric and ferroelectric properties. A-site substitution is also anticipated to significantly alter the macroscopic properties. Therefore, A-site modified three-layered ($Sm_xBi_{4-x}Ti_3O_{12}$, where $x = 0$, 0.37, 0.57, and 0.72 (SBIT)) and four-layered ($SrSm_yBi_{4-y}Ti_4O_{15}$, where $y = 0$, 0.13, 0.43, and 0.78 (SBST15)) compounds were prepared. The intergrowth SBIT-SBST15 was fabricated by mixing the above stated two compounds. The detailed dielectric and ferroelectric properties of these Sm modified BIT-SBT15 intergrowth ceramics were investigated. SBIT and SBST15 were first synthesized independently using the conventional solid-state reaction method. These compounds were then mixed in an agate mortar such that $x + y = 0$, 0.5, 1, and 1.5, calcined at 900 °C for 5 h and then pressed into pellets and sintered at 1100 °C for 5 h. XRD studies revealed the formation of a single phase with the signature of intergrowth oxide. The related strong XRD peaks corresponding to the (118), (200), and (209) planes are found to increase with increasing Sm content. An increase in the full width at half maximum

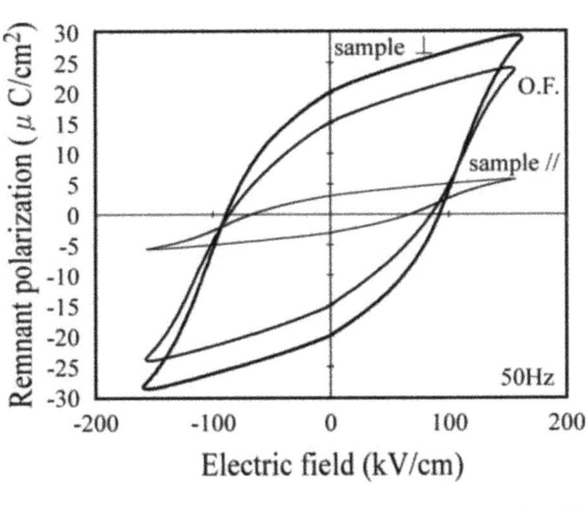

Figure 4.13. *P–E* hysteresis loops of oriented $Bi_{6.5}Sr_{0.5}Ti_4$–$Nb_{0.5}W_{0.5}O_{21}$ ceramic with and without hot-press measured at 50 Hz. [75] 2015, reprinted by permission of the publisher (Taylor and Francis).

value of the strongest XRD peak is attributed to the characteristic nature of the intergrowth. The plate-like morphology was observed by scanning electron microscopic analyses and the grain size was found to decrease with an increase in the Sm concentration in the BST-SBST intergrowth. The orthorhombicity, grain size, and transition temperature were found to decrease with the increase of Sm content. The dielectric and modulus measurements were performed as a function of temperature (RT−600 °C) and frequency (10, 50, and 100 kHz). The transition temperature was found to be around 500 °C. The variations of the imaginary part of the modulus and dielectric constant with the temperature showed a diffuse phase transition. The results are corroborated with a defect-based mechanism and the Curie deviation factor was evaluated using the modified Curie−Weiss law. *P−E* hysteresis loops show a typical narrow hysteresis loop at room temperature for an applied electric field of 70 kV cm^{-1}. However, above room temperature the hysteresis loop area was found to increase. The diffusive nature of the Sm doped intergrowth may be ascribed to the volatile nature of Bi during the sintering process. According to the following Kroger–Vink reaction it can be said that the number of oxygen vacancies increases at higher temperatures:

$$2Bi_{Bi}^{x} + 3O_{o}^{x} \rightarrow 2V_{Bi}^{''} + 3V_{o}^{''} + Bi_2O_3\uparrow \tag{4.8}$$

Based on equation (4.8), it can be said that oxygen vacancy can be compensated by a titanium ion to maintain the charge neutrality condition [77]. The complex defect dipoles play a vital role in the conduction process at lower temperature. But as the temperature is increased, the double ionized vacancies become a single ionized vacancy plus an electron. They start playing a role in the conduction process at higher temperatures. Therefore, substitution of Bi^{3+} with Sm^{3+} ions could improve the ferroelectric properties due to the reduction of oxygen vacancy concentration. This in turn leads to the weakening of the defect mobility and domain pinning effect and results in well-defined hysteresis behaviour at higher temperatures. The results were validated with the tolerance factor values and rotation of TiO_2 octahedra.

4.8 Summary

Therefore, it can be summarized that 1–2, 2–3, and 3–4 intergrowth superlattice structures with a half unit cell of an *m* member structure and another half unit cell of an *m* + 1 member structure can be synthesized. The mismatch between the two parent blocks may be considered as the reason for the improvement in the ferroelectric/piezoelectric properties of these intergrowth Aurivillius phases. Locally disordered structures are found in ordered periodic structures and a small degree of ordering is found in disordered structures. Varying the ratio of constituents in an intergrowth is found to change the ordering in its structure. BTN_2BIT_1 is found to be a disordered structure with local ordering of −23-, −223-, −2223-, and −22-orderings due to kinetic effects. By changing the ratio to BTN_1BIT_2 an order intergrowth with 233 structure was obtained. Single-phase multiferroics in the Aurivillius family, such as Bi_3TiNbO_9–$Bi_3LaTiNbFeO_{12}$ and $Bi_3LaTiNbFeO_{12}$–$Bi_5Ti_3FeO_{15}$, were studied and it was found that the ferroelectric nature of these compounds is enhanced due to intergrowth formation compared to their constituents. However, their magnetic properties remain unaffected. Cobalt (Co) doped $Bi_4NdTi_3Fe_{1-x}Co_xO_{15}$–$Bi_3NdTi_2Fe_{1-x}Co_xO_{12-\delta}$ shows multiferroic behaviour with its magnetodielectric coefficient increasing with an increase in Co content up to a doping concentration of $x = 0.5$. Its ferroelectric properties deteriorate while its ferromagnetic properties improve up to a doping content of $x = 0.25$ and then follows the reverse trend. In most of intergrowths, the ferroelectric properties such as remanent polarization are found to be greater than those of the constituents. Non-successive layer intergrowth as a special case was observed in 1–3 BMO-BIT intergrowth ceramics. The 2–3 intergrowths were found to possess two anomalies in the dielectric versus temperature curves. The peak corresponding to the higher temperature was regarded as the ferroelectric transition temperature. However, in 3–4 intergrowths only one peak was observed in the ε versus T curve which lies between the T_c of its constituents. Doping with various elements such as La, Ce, Er, W, Nb, Sm, and Sr in intergrowths was studied to understand their effect on the structure and properties. It was observed that in A-site doping the dopant initially substitutes the A-site in the perovskite block, which causes enhancement in its ferroelectric nature. On increasing the doping content, the dopant starts occupying the Bi position in the Bi_2O_2 layer, thus weakening its ferroelectric properties. A remarkable increase in the value of remanent polarization was observed in tungsten doped $Bi_4Ti_3O_{12}$–$SrBi_4Ti_4O_{15}$ intergrowth ceramics at a doping concentration of $x = 0.03$. It may be regarded as a potential candidate for commercial applications in FeRAM.

References

[1] Moure A 2018 Review and perspectives of Aurivillius structures as a lead-free piezoelectric system *Appl. Sci.* **8** 62

[2] Yang H, Chen Z, Peng R, Huang H, Fu Z, Zhai X and Lu Y 2018 Superlattice-like structure and enhanced ferroelectric properties of intergrowth Aurivillius oxides *RSC Adv.* **8** 16937–46

[3] Kihlborg L 1978 Order, disorder and defects in intergrowth tungsten bronzes *Chem. Scripta* **14** 79

[4] Hutchison J L, Anderson J S and Rao C N R 1995 Electron microscopy of ferroelectric bismuth oxides containing perovskite layers *Solid State ChemistryWorld Scientific Series in 20th Century Chemistry* (Singapore: World Scientific) pp 66–81

[5] Treacy M M J, Newsam J M and Deem M W 1988 Diffraction from zeolites containing planar faults *MRS Proc.* **138** 497

[6] Kienle L, Babizhetskyy V, Duppel V, Guerin R and Simon A 2007 Lamellar structures in neodymium borides *J. Solid State Chem.* **180** 2775–84

[7] Maresch W V, Czank M and Schreyer W 1994 Growth mechanisms, structural defects and composition of synthetic tremolite: what are the effects on macroscopic properties? *Contrib. Mineral. Petrol.* **118** 297–313

[8] Pring A, Jercher M and Makovicky E 1999 Disorder and compositional variation in the lillianite homologous series *Mineral. Mag.* **63** 917–26

[9] Svensson G 1990 HREM studies of intergrowth between NbO and perovskite in the Ba-, Sr and K-Nb-O systems *Microsc. Microanal. Microstruct.* **1** 343–56

[10] Damjanovic D 1998 Materials for high temperature piezoelectric transducers *Curr. Opin. Solid State Mater. Sci.* **3** 469

[11] Kohn J A and Eckart D W 1964 New hexagonal ferrite, establishing a second structural series *J. Appl. Phys.* **35** 968–9

[12] Kohn J A, Eckart D W and Cook C F Jr 1967 Determination of large crystal structures by combined x-ray diffraction and electron microscopy *Mater. Res. Bull.* **2** 55–68

[13] Rao C N R 1985 Intergrowth structures in inorganic solids: a new class of materials *Bull. Mater. Sci.* **7** 155–78

[14] Gao X, Gu H, Li Y X, Yi Z G, Čeh M and Žagar K 2011 Structural evolution of the intergrowth bismuth-layered $Bi_7Ti_4NbO_{21}$ *J. Mater. Sci.* **46** 5423–31

[15] Shibuya A, Ikemori S, Wu W B, Noda M and Okuyama M 2003 Preparation and basic properties of ferroelectric thin films having a superlattice structure of 2 Bi_3TiNbO_9 units–1 $Bi_4Ti_3O_{12}$ unit *Appl. Phys. Lett.* **83** 1411–3

[16] Boullay P, Trolliard G, Mercurio D, Perez-Mato J M and Elcoro L 2002 Toward a unified approach to the crystal chemistry of Aurivillius-type compounds: I. The structural model *J. Solid State Chem.* **164** 252–60

[17] Zhang F, Li Y, Gu H and Gao X 2014 Local orderings in long-range-disordered bismuth-layered intergrowth structure *J. Solid State Chem.* **212** 165–70

[18] Zhang F, Wahyudi O, Liu Z, Gu H and Li Y 2018 Preparation and electrical properties of a new-type intergrowth bismuth layer-structured $(Bi_3TiNbO_9)_1(Bi_4Ti_3O_{12})_2$ ceramics *J. Alloys Compd.* **753** 54–9

[19] Luo S, Noguchi Y, Miyayama M and Kudo T 2001 Rietveld analysis and dielectric properties of Bi_2WO_6–$Bi_4Ti_3O_{12}$ ferroelectric system *Mater. Res. Bull.* **36** 531–40

[20] Shibuya A, Noda M, Okuyama M and Saito K 2006 Preparation and characterization of natural-superlattice-structured Bi_2MoO_6–$Bi_4Ti_3O_{12}$ (m = 1–3) thin films *Integr. Ferroelectr.* **79** 15–23

[21] Zhang D L, Huang W C, Chen Z W, Zhao W B, Feng L, Li M, Yin Y W, Dong S N and Li X G 2017 Structure evolution and multiferroic properties in cobalt doped $Bi_4NdTi_3Fe_{1-x}Co_xO_{15}$–$Bi_3NdTi_2Fe_{1-x}Co_xO_{12-\delta}$ intergrowth Aurivillius compounds *Sci. Rep.* **7** 43540

[22] Mao X, Wang W, Chen X and Lu Y 2009 Multiferroic properties of layer-structured $Bi_5Fe_{0.5}Co_{0.5}Ti_3O_{15}$ ceramics *Appl. Phys. Lett.* **95** 082901

[23] Kubota M, Oka K, Yabuta H, Miura K and Azuma M 2013 Structure and magnetic properties of $BiFe_{1-x}Co_xO_3$ and $Bi_{0.9}Sm_{0.1}Fe_{1-x}Co_xO_3$ *Inorg. Chem.* **52** 10698–704

[24] Kobayashi T, Noguchi Y and Miyayama M 2005 Enhanced spontaneous polarization in superlattice-structured $Bi_4Ti_3O_{12}$–$BaBi_4Ti_4O_{15}$ single crystals *Appl. Phys. Lett.* **86** 012907

[25] Yang F J, Su P, Wei C, Chen X Q, Yang C P and Cao W Q 2011 Large magnetic response in $(Bi_4Nd)Ti_3(Fe_{0.5}Co_{0.5})O_{15}$ ceramic at room-temperature *J. Appl. Phys.* **110** 126102

[26] Chen X, Xiao J, Yao J, Kang Z, Yang F and Zeng X 2014 Room temperature magneto-electric coupling study in multiferroic $Bi_4NdTi_3Fe_{0.7}Ni_{0.3}O_{15}$ prepared by a multicalcination procedure *Ceram. Int.* **40** 6815–9

[27] Tilley D R and Scott J F 1982 Frequency dependence of magnetoelectric phenomena in $BaMnF_4$ *Phys. Rev.* B **25** 3251

[28] Fiebig M 2005 Revival of the magnetoelectric effect *J. Phys. D: Appl. Phys.* **38** R123

[29] Kikuchi T, Watanabe A and Uchida K 1977 A family of mixed-layer type bismuth compounds *Mater. Res. Bull.* **12** 299–304

[30] Zhang H, Yan H and Reece M J 2010 High temperature lead-free relaxor ferroelectric: intergrowth Aurivillius phase $BaBi_2Nb_2O_9$–$Bi_4Ti_3O_{12}$ ceramics *J. Appl. Phys.* **107** 104111

[31] Noguchi Y, Miyayama M and Kudo T 2000 Ferroelectric properties of intergrowth $Bi_4Ti_3O_{12}$–$SrBi_4Ti_4O_{15}$ ceramics *Appl. Phys. Lett.* **77** 3639–41

[32] Shimakawa Y, Kubo Y, Nakagawa Y, Kamiyama T, Asano H and Izumi F 1999 Crystal structures and ferroelectric properties of $SrBi_2Ta_2O_9$ and $Sr_{0.8}Bi_{2.2}Ta_2O_9$ *Appl. Phys. Lett.* **74** 1904–6

[33] Yi Z G, Wang Y, Li Y X and Yin Q R 2006 Ferroelectricity in intergrowth Bi_3TiNbO_9–$Bi_4Ti_3O_{12}$ ceramics *J. Appl. Phys.* **99** 114101

[34] Boullay P and Mercurio D 2004 Structural behaviour of the mixed-layer Aurivillius-phase $Bi_7Ti_4NbO_{21}$ *Integr. Ferroelectr.* **62** 149–54

[35] Boullay P, Trolliard G, Mercurio D, Perez-Mato J M and Elcoro L 2002 Toward a unified approach to the crystal chemistry of Aurivillius-type compounds: II. $Bi_7Ti_4NbO_{21}$, a case study *J. Solid State Chem.* **164** 261–71

[36] Thompson J G, Rae A D, Withers R L and Craig D C 1991 Revised structure of Bi_3TiNbO_9 *Acta Crystallogr.* B **47** 174–80

[37] Rae A D, Thompson J G and Withers R L 1991 Structure refinement of commensurately modulated bismuth tungstate, Bi_2WO_6 *Acta Crystallogr.* B **47** 870–81

[38] Machura D, Surowiak Z, Rymarczyk J and Ilczuk J 2008 Mixed layer-structured ferroelectrics on the basis of $Bi_4Ti_3O_{12}$ *Eur. Phys. J. Spec. Top.* **154** 135–8

[39] Parida G and Bera J 2014 Dielectric properties of $Bi_4Ti_3O_{12}$–$SrBi_4Ti_4O_{15}$ intergrowth ceramics synthesized by a modified oxalate route *Phase Transit.* **87** 452–9

[40] Tian X, Qu S, Wang B, Du H and Xu Z 2014 Intergrowth bismuth layer-structured $Na_{0.5}Bi_{2.5}Nb_2O_9$–$Bi_4Ti_3O_{12}$ high temperature ferroelectrics ceramics *J. Inorg. Organomet. Polym. Mater.* **24** 355–9

[41] Zhu J S, Su D, Lu X M, Qin H X, Wang Y N, Wang D Y, Chan H L W, Wong K H and Choy C L 2002 La-doped effect on the ferroelectric properties of $Bi_4Ti_3O_{12}$–$SrBi_4Ti_4O_{15}$ thin film fabricated by pulsed laser deposition *J. Appl. Phys.* **92** 5420–4

[42] Xue C, Sun X, Zhang Y, Zhao Y, Zhu H, Yang Q, Liu M, Wang C and Ouyang J 2017 Natural-superlattice structured $CaBi_2Nb_2O_9$–$Bi_4Ti_3O_{12}$ ferroelectric thin films *Ceram. Int.* **43** 8459–65

[43] Zhang Y, Ouyang J, Zhang J, Li Y, Cheng H, Xu H, Liu M, Cao Z P and Wang C M 2016 Strain engineered $CaBi_2Nb_2O_9$ thin films with enhanced electrical properties *ACS Appl. Mater. Interf.* **8** 16744–51

[44] Barna A, Pécz B and Menyhard M 1999 TEM sample preparation by ion milling/amorphization *Micron* **30** 267–76

[45] McCaffrey J P, Phaneuf M W and Madsen L D 2001 Surface damage formation during ion-beam thinning of samples for transmission electron microscopy *Ultramicroscopy* **87** 97–104

[46] Chu F, Damjanovic D, Steiner O and Setter N 1995 Piezoelectricity and phase transitions of the mixed-layer bismuth titanate niobate $Bi_7Ti_4NbO_{21}$ *J. Am. Ceram. Soc.* **78** 3142–4

[47] Miyabayashi H, Hiruma Y, Nagata H and Takenaka T 2010 Temperature dependences of piezoelectric properties of grain-oriented $SrBi_2Nb_2O_9$–$Bi_4Ti_3O_{12}$ ceramics *Ferroelectrics* **404** 147–51

[48] Kharitonova E P and Voronkova V I 2007 Synthesis and electrical properties of mixed-layer Aurivillius phases *Inorg. Mater.* **43** 1340–4

[49] Yan J and Hu G D 2018 Enhanced ferro- and piezoelectric properties of $Bi_4Ti_3O_{12}$–$CaBi_4Ti_4O_{15}$ thin film on Pt(111)/Ti/SiO_2/Si substrate *Mater. Res. Express* **5** 056306

[50] Noguchi Y and Miyayama M 2001 Large remanent polarization of vanadium-doped $Bi_4Ti_3O_{12}$ *Appl. Phys. Lett.* **78** 1903–5
[51] Hu G D 2006 Orientation dependence of ferroelectric and piezoelectric properties of $Bi_{3.15}Nd_{0.85}Ti_3O_{12}$ thin films on Pt (100)/TiO_2/SiO_2/Si substrates *J. Appl. Phys.* **100** 096109
[52] Rae A D, Thompson J G, Withers R L and Willis A C 1990 Structure refinement of commensurately modulated bismuth titanate, $Bi_4Ti_3O_{12}$ *Acta Crystallogr.* B **46** 474–87
[53] Yi Z G, Li Y X, Wang Y and Yin Q R 2006 Dielectric and ferroelectric properties of intergrowth $Bi_{7-x}La_xTi_4NbO_{21}$ ceramics *Appl. Phys. Lett.* **88** 152909
[54] Mercurio D, Trolliard G, Hansen T and Mercurio J P 2000 Crystal structure of the ferroelectric mixed Aurivillius phase $Bi_7Ti_4NbO_{21}$ *Int. J. Inorg. Mater.* **2** 397–406
[55] Newnham R E, Wolfe R W and Dorrian J F 1971 Structural basis of ferroelectricity in the bismuth titanate family *Mater. Res. Bull.* **6** 1029–39
[56] Yi Z G, Li Y X, Zeng J T, Yang Q B, Wang D, Lu Y Q and Yin Q R 2005 Lanthanum distribution and dielectric properties of intergrowth $Bi_{5-x}La_xTiNbWO_{15}$ ferroelectrics *Appl. Phys. Lett.* **87** 202901
[57] Zhu J, He J H and Chen X B 2006 Substitution site of lanthanum ions in La-doped $SrBi_4Ti_4O_{15}$ and $Bi_4Ti_3O_{12}$–$SrBi_4Ti_4O_{15}$ *Integr. Ferroelectr.* **79** 265–71
[58] Zhu J, Chen X B, Lu W P, Mao X Y and Hui R 2003 Properties of lanthanum-doped $Bi_4Ti_3O_{12}$–$SrBi_4Ti_4O_{15}$ intergrowth ferroelectrics *Appl. Phys. Lett.* **83** 1818–20
[59] Zhu J, Mao X Y and Chen X B 2005 Substitution site of La ions in La-doped $Bi_4Ti_3O_{12}$–$SrBi_4Ti_4O_{15}$ intergrowth ferroelectrics *J. Cryst. Growth* **277** 462–6
[60] Zhang S T, Xiao C S, Fang A A, Yang B, Sun B, Chen Y F, Liu Z G and Ming N B 2000 Ferroelectric properties of $Sr_2Bi_4Ti_5O_{18}$ thin films *Appl. Phys. Lett.* **76** 3112–4
[61] De Araujo C P, Cuchiaro J D, McMillan L D, Scott M C and Scott J F 1995 Fatigue-free ferroelectric capacitors with platinum electrodes *Nature* **374** 627–9
[62] Kim S K, Miyayama M and Yanagida H 1996 Electrical anisotropy and a plausible explanation for dielectric anomaly of $Bi_4Ti_3O_{12}$ single crystal *Mater. Res. Bull.* **31** 121–31
[63] Liu F, Jiang X, Chen C, Nie X, Huang X, Chen Y, Hu H and Su C 2019 Structural, electrical and photoluminescence properties of Er^{3+}-doped $SrBi_4Ti_4O_{15}$–$Bi_4Ti_3O_{12}$ intergrowth ceramics *Front. Mater. Sci.* **13** 99–106
[64] Parida G and Bera J 2014 Electrical properties of niobium doped $Bi_4Ti_3O_{12}$–$SrBi_4Ti_4O_{15}$ intergrowth ferroelectrics *Ceram. Int.* **40** 3139–44
[65] Jiang Y, Jiang X, Chen C, Chen Y, Jiang X and Tu N 2017 Photoluminescence and electrical properties of Er^{3+}-doped $Na_{0.5}Bi_{4.5}Ti_4O_{15}$–$Bi_4Ti_3O_{12}$ inter-growth ferroelectric ceramics *Front. Mater. Sci.* **11** 51–8
[66] Castro A, Millan P and Jimenez B 1998 Influence of Bi-site substitution on the ferroelectricity of the Aurivillius compound $Bi_2SrNb_2O_9$ *J. Mater. Res.* **13** 2565–71
[67] Wu Y, Limmer S J, Chou T P, Nguyen C and Cao G 2002 Influence of tungsten doping on dielectric properties of strontium bismuth niobate ferroelectric ceramics *J. Mater. Sci. Lett.* **21** 947–9
[68] Kumar S and Varma K B R 2009 Influence of lanthanum doping on the dielectric, ferroelectric and relaxor behaviour of barium bismuth titanate ceramics *J. Phys. D: Appl. Phys.* **42** 075405
[69] Jiang Y, Jiang X, Chen C, Chen Y, Jiang X, Tu N, Xia X, Luo Y and Zhu S 2017 Structural and electrical properties of La^{3+}-doped $Na_{0.5}Bi_{4.5}Ti_4O_{15}$–$Bi_4Ti_3O_{12}$ inter-growth high temperature piezoceramics *Ceram. Int.* **43** 6446–52

[70] Rehman F, Jin H B and Li J B 2016 Effect of reduction/oxidation annealing on the dielectric relaxation and electrical properties of Aurivillius $Na_{0.5}Gd_{0.5}Bi_4Ti_4O_{15}$ ceramics *RSC Adv.* **6** 35102–9

[71] Zhu J, Hui R, Mao X Y, Lu W P, Chen X B and Zhang Z P 2003 Properties and lanthanum distribution of $Bi_{4-x}La_xTi_3O_{12}$–$SrBi_{4-y}La_yTi_4O_{15}$ intergrowth ferroelectrics *J. Appl. Phys.* **94** 5143–6

[72] Takenaka T and Sakata K 1981 Electrical properties of grain-oriented ferroelectric ceramics in some lanthanum modified layer-structure oxides *Ferroelectrics* **38** 769–72

[73] Wang X, Jiang X, Nie X, Huang X, Chen C, Wang J, Xu J and Huang L 2021 Piezoelectric properties of cerium-doped bismuth barium titanate intergrowth ceramics *Process. Appl. Ceram.* **15** 120–7

[74] Wang W, Zhu J, Mao X Y and Chen X B 2007 Properties of tungsten-doped $Bi_4Ti_3O_{12}$–$SrBi_4Ti_4O_{15}$ intergrowth ferroelectrics *Mater. Res. Bull.* **42** 274–80

[75] Ogawa H, Iida D, Takahashi S, Moriyama T and Kan A 2016 Ferroelectric properties and crystal structure of mixed-layer $Bi_{7-x}Sr_xTi_4Nb_{1-x}W_xO_{21}$ ceramics *Ferroelectrics* **498** 1–11

[76] Shobhan B B, Narendra B S, Prasad G, Kumar G S and Prasad N V 2020 Structure and dielectric properties of Sm^{3+} modified $Bi_4Ti_3O_{12}$–$SrBi_4Ti_4O_{15}$ intergrowth ferroelectrics *Process. Appl. Ceram.* **14** 260–7

[77] Afqir M, Tachafine A, Fasquelle D, Elaatmani M, Carru J C, Zegzouti A, Daoud M, Sayouri S, Lamcharfi T D and Zouhairi M 2018 Structural, electric and dielectric properties of Eu-doped $SrBi_2Nb_2O_9$ ceramics obtained by co-precipitation route *Process. Appl. Ceram.* **12** 72–7

Chapter 5

Applications and future perspectives of Aurivillius oxides

The family of Aurivillius oxides has lately gained significance as piezoelectric ceramics due to the better stability, high working temperature, and higher operating frequency of these materials. These ceramics are used mostly in ferroelectric random-access memory (FeRAM), resistive random-access memory (RRAM), piezoelectric resonator devices which need to exhibit a very stable resonant frequency and possess electro-optical applications, piezoelectric nanogenerators, piezoelectric sensors, multi-layer ceramic capacitors, transducers, micro-electro-mechanical systems (MEMS), optoelectronic devices, and gas sensors [1–3]. This chapter deals with the preparation procedures and performance of devices or prototypes using Aurivillius phase oxide materials for some of the above-mentioned applications.

5.1 Ferroelectric random-access memory (FeRAM)

The working principle of ferroelectric random-access memory (FeRAM) is based on a ferroelectric capacitor consisting of a thin layer of ferroelectric material [4–6]. The external field dependent potential difference developed across the ferroelectric capacitor is in either the positive or negative direction. This implies that the polarization in the ferroelectric film may be bidirectional. After the external potential is reset to zero, the remanent polarization in the ferroelectric film induces an opposite charge on the capacitor electrodes. The interaction between the remanent polarization and the induced opposite charge gives rise to a steady state in which the remanent polarization may be maintained for a longer period of time. This feature may be utilized to define the zero and one states for memory applications. The structure of an FeRAM memory cell is composed of an NMOS transistor and a ferroelectric capacitor that are used as a switch and data storage, respectively.

Stacked ferroelectric thin film capacitors of Aurivillius phase $SrBi_4Ti_4O_{15}$/$CaBi_4Ti_4O_{15}$ (SBT15/CBT15), and perovskite phase $Ba_{0.5}Sr_{0.5}TiO_3$ (BST) were

doi:10.1088/978-0-7503-3419-8ch5

fabricated by Noda *et al* [7]. They fabricated a capacitor with SBT15 (or CBT15) film having a structure of Au/SBT15 (or CBT15)/Pt(100)/$SrTiO_3$(100) and also fabricated a BST film capacitor with Au/BST/Pt(111)/Ti/SiO_2/Si(100) for reference. The film thicknesses of SBT15/CBT15 and BST were around 500 and 750 nm, respectively. The top and bottom electrodes in both capacitors were made up of Au (about 700 nm) and Pt (about 300 nm), respectively. The capacitance was measured between the two nearest top Au circular electrodes of diameter 200 μm.

Compared to the performance of the perovskite barium titanate family of $Ba_{0.5}Sr_{0.5}TiO_3$, the SBT15 film capacitor possesses low leakage ~10^{-7} A cm^{-2} at 250 kV cm^{-1}. This was lower by an order of magnitude than that for the BST film, even for thinner SBT15 films. The temperature coefficient of capacitance (TCC) of the SBT15 or CBT15 film was found to be ~100–250 ppm K^{-1} and the dielectric loss was found to be as low as 0.02 measured at a frequency of 1 MHz, which was much lower than that for BST film. As the SBT15 and CBT15 films possess opposite polarities of TCC in the experimental observations by Noda *et al*, they might cancel each other out in the SBT15/CBT15 composite capacitor. Therefore, the bismuth layer structured dielectric (BLSD) films of SBT15 and CBT15 may be regarded as more effective for applications in high temperature and high permittivity capacitors compared to the currently used barium perovskite oxide family for various device applications. While working at different temperatures and electric fields, SBT15 and CBT15 film capacitors are expected to show less change in the values of capacitance than the conventional and commercial perovskite capacitors.

5.2 Resistive random-access memory (RRAM)

A Pt/Bi_2MoO_6/FTO device was fabricated with a characteristic bipolar resistive switching behaviour by Guo *et al* [8]. They fabricated the device with the help of a solution prepared by dissolving $Bi(NO_3)_3$ in potassium iodide (KI) with the pH maintained at 7 using HNO_3. This solution was mixed with p-benzoquinone absolute ethanol solution, and was stirred continuously until homogenized. Fluorine doped tin oxide (FTO) substrates of area 1–2 cm^2 were cleaned by sonication in diluted detergent, acetone, and ethanol for 30 min. The electrodes were deposited using an electrochemical workstation in a three electrode system using the FTO substrate as the working electrode (WE), a platinum wire as the counter electrode (CE), and Ag/AgCl as the reference electrode (RE). Cathode deposition was carried out potentiostatically at room temperature (RT). The as-prepared BiOI films were removed using deionized (DI) water, and then dried in an oven at 100 °C. Then $(NH_4)_6Mo_7O_{24}$–$4H_2O$ dissolved in dimethyl sulfoxide was drop-casted onto the BiOI film and the obtained films were heated in a muffle furnace at 500°C for 4 h at a heating rate of 2 °C min^{-1}. The excess of MoO_3 on the films was removed by immersing them in 0.5 M NaOH for 30 s. The obtained pure Bi_2MoO_6 films were washed with ultrapure water and dried at RT. Finally, 100 nm thick Pt top electrodes with a diameter of 200 mm were deposited by sputtering a Pt target in a pure argon atmosphere using a direct current magnetron sputter with a shadow mask covered on the Bi_2MoO_6 film. The RRAM performance was evaluated in the dark under ambient conditions using a two-probe method with a Keithley 2400 semiconductor parameter analyzer. During the electrical characterization of the samples, the bottom electrode was connected to the ground potential. Each cell was then characterized using an impedance analyzer at the desired cell status following the resistive switching by the semiconductor parameter analyzer. The amplitude and frequency for the impedance measurements were maintained at 1 V and 100 Hz–1 MHz, respectively. A schematic diagram of the vertical stack structure of the Pt/Bi_2MoO_6/FTO is shown in figure 5.1(a). An SEM image of the cross-sectional view of the device structure is shown in figure 5.1(b). The *I*–*V* cycle was studied for around 500 cycles with high stability, exceptional data retention, and the potential for multilevel data storage. Impedance and dielectric spectroscopy were performed to evaluate the features of both the high resistance state (HRS) and the low resistance state (LRS). They were found to show two distinct characteristics in the impedance spectra and a large change in the loss tangent of approximately three orders of magnitude. The activation energy and relaxation time were evaluated from the dielectric relaxation peak and were found to be 0.21 eV and 6.9×10^{-8} s, respectively. The orientation of the oxygen vacancies along the direction of the applied alternating electric field gives rise to the polarization in the filament. A model involving the creation and separation of conductive filaments due to oxygen ion movement was proposed to understand the resistive switching behaviour of the device. This process of device fabrication and understanding its conduction mechanism in terms of the oxygen ion conduction in Bi_2MoO_6 for non-volatile memory devices provides an approach to investigate the course of oxygen vacancy movement in the filaments of valence change memory devices.

Multiferroic and ferroelectric $Bi_{0.8}Pr_{0.2}Fe_{0.95}Mn_{0.05}O_3$ (BPFMO)/$Bi_{3.96}Gd_{0.04}Ti_{2.95}W_{0.05}O_{12}$ (BGTWO) bilayer thin films were fabricated on Pt(111)/Ti/SiO_2/Si (100) substrates by Kao *et al* [9]. A simple method was used for depositing crack-free BPFMO/BGTWO bilayer thin film. The precursor solution of BPFMO and BGTWO was prepared using the sol–gel technique. The BGTWO solution was spin-coated on the Pt/Ti/SiO_2/Si(100) substrate. The spin coating was performed 15 times and after each coating step the gel films were pyrolyzed on a hot plate at 300 °C for 2 min before

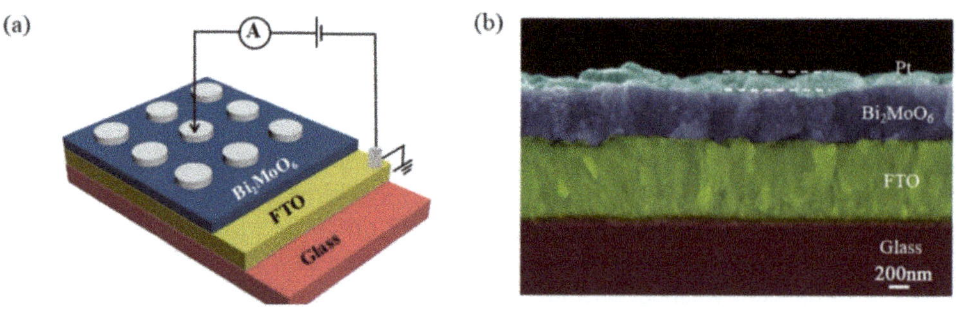

Figure 5.1. (a) Schematic of the vertical stack structure of Pt/Bi_2MoO_6/FTO. (b) SEM image of the cross-sectional view of the device structure. Reproduced from [8] with permission of The Royal Society of Chemistry.

final annealing. After multiple coatings these thin films were heat treated at 650 °C for 2 min using rapid thermal annealing (RTA) at a heating rate of 500 °C min^{-1} in oxygen atmosphere. The required thin film thickness was attained by repeated spin coating and annealing cycles. The BPFMO solution was then spin-cast six times on BGTWO/Pt/Ti/SiO_2/Si(100) substrates using the same process. Finally, the BPFMO/BGTWO bilayer thin films were annealed at 450 °C for 2 min by RTA at a heating rate of 400 °C min^{-1} in oxygen atmosphere. The leakage current density of the bilayer thin films is reduced significantly. In addition, the BPFMO/BGTWO bilayer thin films displayed improved ferroelectric and ferromagnetic properties in comparison to that of pure BPFMO thin film, with a remnant polarization of 62 μC cm^{-2} and a remnant magnetization of 4.6 emu g^{-1}. The electrical conduction behaviour of BPFMO/BGTWO bilayer thin film RRAM for the set/reset state was demonstrated, and it showed the hopping conduction and Schottky emission mechanisms. Figure 5.2 depicts the bipolar *I–V* switching curves of the BPFMO/BGTWO bilayer thin films for 100 cycles. The inset of figure 5.2 shows that when a voltage of 3 V was applied to the initial forming process, the compliance current was 10 mA. Above the applied positive voltage for the RRAM devices, low current is abruptly shifted to high current, which was denoted by the set state. To describe the reset state, the high current of the RRAM devices was shifted to low current. The low resistive status (LRS) exhibited the Schottky and hopping conduction mechanisms in the *I–V* fitting curves. For the high resistive status (HRS), the RRAM devices showed the hopping conduction mechanism. Therefore the BPFMO/BGTWO bilayer RRAM devices showed a diverse electrical conduction mechanism in the reset state. This electrical hopping conduction effect may be due to the electrons being transferred to the metal-ion cluster of the multiferroic/ferroelectric composite.

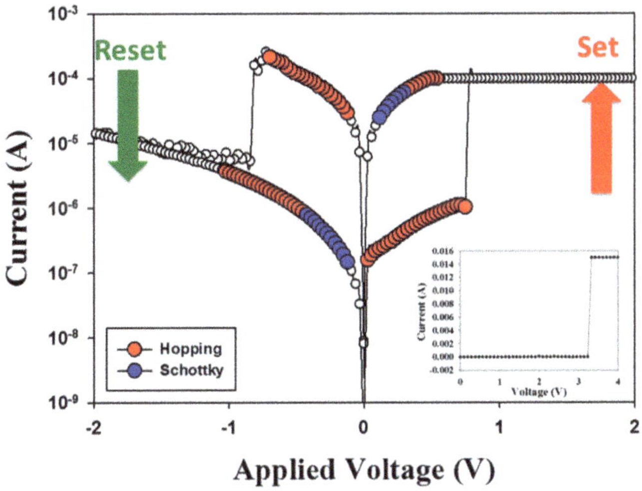

Figure 5.2. Bipolar *I–V* switching curves of BPFMO/BGTWO bilayer thin films. Reproduced from [9] with permission from MDPI.

5.3 Acoustic piezoelectric sensor

Bismuth titanate ($Bi_4Ti_3O_{12}$), a material with a high Curie temperature ($T_c = 675$ °C), high electrical resistivity at 500 °C, and stable dielectric properties as a function of temperature, was prepared using the mixed oxide route and was used for acoustic sensor fabrication by Megriche *et al* [10]. A piezoelectric sensor prototype was fabricated by depositing a thin film of ceramic on a foil. However, a temperature limit was observed in such a device due to the non-binding of the ceramic and metal. The viability of such a thick film sensor was documented. The dielectric constant and resistivity coefficients were found to be stable up to 450 °C for the as-prepared thick film sensor prototype. For this sensor type, bismuth titanate was deposited on a SiC substrate. A pressure sensor built like a telephonic buzzer was designed and fabricated. It was made up of a thin disc of piezoelectric material of thickness 300 μm and diameter 22 mm deposited on a 200 mm thick brass foil of diameter 30 mm. This flexible device gains more significance when fixed on its ends. As acoustic waves strike the unimorph, the generation of electric charges takes place according to the direct piezoelectric effect. The amount of generated charges corresponds to the applied pressure. The buzzer produces a signal with a frequency of 2 kHz that can be detected by sensor placed at a distance without further amplification. Figure 5.3 shows the fabricated acoustic sensor device. A linear dependence of output voltage on the input voltage from 1–16 V of the sensor was observed. The sensitivity of the sensor is the ratio of output voltage and the applied pressure, the pressure applied at the buzzer being related to the input voltage. Subsequently, it was reported that this sensitivity remains constant in the pressure range (up to 100 mPa) of the buzzer used in the experiment performed by Megriche *et al.* However, this device could not be used above a temperature of 300 °C because the binder used to mount the disk does not work beyond that temperature. To overcome this issue, one more type of sensor was fabricated with a layer of the substituted $Bi_4Ti_3O_{12}$ composition on an alumina substrate having dimensions of $25 \times 12 \times 0.25$ mm^3. The post-annealed thickness of the film was 60 μm. The dielectric properties were found to be stable with temperature for this sensor. Similar values of piezoelectric measurements were obtained for the

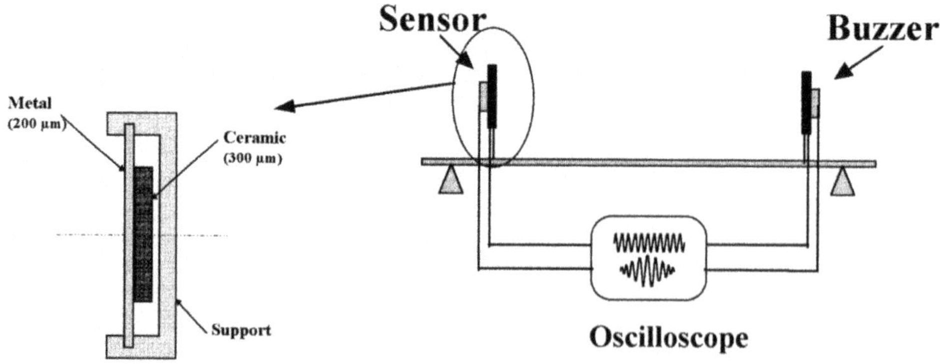

Figure 5.3. Prototype of an acoustic pressure sensor. Reproduced from [10]. Copyright (1999), with permission from Elsevier.

thick film sensors as that of its ceramic counterparts. The sensing ability of the thick film BIT sensor could not be evaluated due to its non-flexibility, as the thick film was deposited on an alumina substrate.

5.4 Biosensors

Glucose oxidase (GOx) based glucose biosensors have been used widely in clinical diagnosis until recently. However, compared to the enzyme imitators, the catalytic efficacy of enzyme based glucose biosensors needs to be improved [11, 12]. Bismuth titanate ($Bi_4Ti_3O_{12}$), being an n-type semiconductor with an energy gap (E_g) = 2.84 eV, has come into focus due to its exceptional crystalline and electronic properties [13–15]. Also, creating a heterojunction is the most suitable way to improve the catalytic activity. When the p-type and n-type semiconductors form a heterojunction, the internally developed electric field increases the charge carrier mobility and electron–hole pair creation ability [16]. NiO, as a p-type semiconductor with an energy gap of 3.5 eV, has been investigated widely due to its high catalytic activity, hole mobility, biocompatibility, and environmentally friendly nature [17–19]. Thus Mao *et al* [20] synthesized a GOx/p-NiO/n-$Bi_4Ti_3O_{12}$ sandwich heterojunction that can be used in GOx based glucose biosensors. Dandelion-like $Bi_4Ti_3O_{12}$ nanowires were grown directly on titanium foil substrates by a facile hydrothermal method and a compact layer of NiO was deposited by PLD after immobilization of GOx. Enzyme immobilization was carried out by dipping the as-prepared $Bi_4Ti_3O_{12}$/titanium foil electrode in 0.1 M phosphate buffer solution (PBS, pH 7.4) with 10 mg mL^{-1} of GOx with periodic shaking. Then the GOx/$Bi_4Ti_3O_{12}$/Ti foil was dried at 4 °C overnight in a refrigerator, followed by washing to remove the loose GOx. As the hydrothermal method does not require the use of binders, a binder-free p–n sensor is a noteworthy innovation that achieves greater direct electrochemical properties. The GOx/p-NiO/n-$Bi_4Ti_3O_{12}$ heterojunction biosensor displays a remarkably improved glucose sensitivity, small response time, low detection limit, and good selectivity. An intimate contact between $Bi_4Ti_3O_{12}$ nanowires and titanium substrate is developed due to the direct growth of nanowires on Ti substrate. The dandelion-like spherical $Bi_4Ti_3O_{12}$ bunches offer a larger sensitive interface between GOx and glucose, consequently enhancing the electrochemical sensing performance. The explicit sandwich structure is responsible for facilitating the transfer of electrons between the GOx active centres and electrode surface. In addition, the higher oxidation current density of the GOx/p-NiO/n-$Bi_4Ti_3O_{12}$ biosensor can be ascribed to the synergetic catalysis of a GOx and p-NiO/n-$Bi_4Ti_3O_{12}$ heterojunction. This study lays down a roadmap for the much required nanostructure controlled properties of the GOx based glucose biosensor with excessively higher sensitivity, which overcomes the constraints of the enzyme based glucose sensors and can be used for potential practical applications.

5.5 Pressure sensors

Pressure sensors with nanocrystalline bismuth titanate ($Bi_4Ti_3O_{12}$) thin film were fabricated using the sol–gel process by Wei *et al* [21]. The BIT solution prepared using the sol–gel technique was coated using spin coater on $Si/SiO_2/RuO_2/TiO_2$ substrates. A 75 nm thick buffer layer film of TiO_2 was deposited by the electron beam evaporation method to avoid atomic inter-diffusion between the bismuth titanate film and the base substrate. The layer-by-layer coating of BIT on the substrate was carried out at an RPM of 3300 for 30 s, followed by heating of each layer at 300 °C in air for 15 min to evaporate the organic solvents. This process of coating and heating the film was carried out four times and finally the film was annealed at 450 °C for 60 s for crystallization and densification. Aluminium film was deposited as the top electrode. The electromechanical response of the pressure sensor was measured using a pneumatic loading method. The pneumatic measurement system comprised three parts: an air compressor system, a pressure vessel, and the electrometer as a data storage system. The pressure was measured in the range 17.5 psi to 65 psi at room temperature. The air compressor was used for compression of air and was impinged at the sensor. The pressure vessel was used to mount the sensor inside. The voltage generated by the sensor was measured by the electrometer. It was observed that the as-prepared pressure sensor was sensitive to the applied pressure and the response was recorded when the pressure was removed from the chamber. The measurements were recorded for three cycles for each of the pressure loads. It was observed that the sensor response was reproducible for the number of measurement cycles. A linear response between the applied pressure loads of 17.5–65 psi was attained for the as-prepared $Bi_4Ti_3O_{12}$ pressure sensor, where the value obtained for the correlation coefficient, R^2, was 0.9813. The voltage sensitivity of the sensor was determined from the slope of the graph of applied pressure versus voltage and the reported value was 0.3 mV psi^{-1}.

5.6 Piezoelectric nanogenerator

Devices that convert mechanical energy into electric energy are called nanogenerators. Enhancing the electrical productivity of nanogenerators is necessary for them to be used in applications for powering various electronic devices. An alternative method of improving the power output is to use ferroelectric materials with a higher piezoelectric coefficient and electromechanical conversion efficiency [22–24]. Bismuth titanate, $Bi_4Ti_3O_{12}$ (BIT), belonging to the Aurivillius family is a potential material for environmentally friendly piezoelectric devices due to their high piezoelectricity, electro-optical properties, ferroelectricity, ionic conductivity, low sintering temperature, and high Curie temperature [25]. A bismuth titanate–PDMS nanocomposite nanogenerator was fabricated by Abinnas *et al* [26] to harvest mechanical energy from finger tapping movements. The bismuth titanate was synthesized using the wet chemical route. The phase analysis, morphology, and properties of the obtained samples were studied using XRD, FESEM, TEM, FT-Raman studies, and dielectric studies. A paste of PDMS was made using chloroform. The calcined BIT powder sample was mixed with PDMS paste to obtain BIT-PDMS nanocomposite paste. The nanocomposite paste was carefully mixed to form a homogeneous mixture. The BIT-PDMS film was deposited on an ITO by spin coating the nanocomposite paste at an rpm of 3000, which was then heat treated in an oven. The top electrode of the nanocomposite film was another ITO substrate placed over it. Therefore, a BIT-PDMS nanocomposite film sandwiched between two ITO electrodes forms a simple nanogenerator. The copper wires were attached to the top and bottom ITO electrodes with the help of conducting epoxy paste, which can be used for measuring the output potential difference developed. To improve the device performance the device was poled by an application of an external electric field of 5 kV for 3 h. Ferroelectric properties of the as-prepared bismuth titanate samples were evaluated using the *P–E* loop. A well saturated hysteresis loop was obtained which is a characteristic of ferroelectric material. A schematic diagram of the BIT-PDMS nanocomposite nanogenerator is shown in figure 5.4. PDMS assists in providing flexibility to the nanogenerator and also contributes in generating piezoelectric potential. Poling is carried out during the fabrication of the nanogenerators. The poling electric field stimulates the BIT-PDMS nanocomposite to produce dipole moments that become aligned in the direction of the applied field. This process expands the material along the direction of the poling axis and contracts it in the directions perpendicular to it. This device generated an output voltage of 0.8 V by gathering the energy released by the tapping of a finger. The output from such a nanocomposite nanogenerator can be enhanced to give an improved performance by changing the combination ratio of BIT and PDMS. This device has the advantage of ease of fabrication and is eco-friendly. This would facilitate the mass production of this type of nanogenerator.

A multiple unit hybrid piezo-triboelectric nanogenerator (HNG) device was fabricated from $Bi_4Ti_3O_{12}$ (BIT)–polydimethylsiloxane (PDMS) composite films and its sensing ability was compared to the individual PENG and TENG based devices by Hajra *et al* [27]. The BIT sample consisting of orthorhombic crystal structure with a space group of *Aba2* was prepared using the conventional solid state reaction route. It showed colossal dielectric properties and non-Debye type relaxation. The origin of the colossal dielectric properties may be ascribed to the cumulative effects of interfacial polarization, hopping polarization, and extrinsic electrode effects. The coexistence of ferroelectric and ferromagnetic behaviour

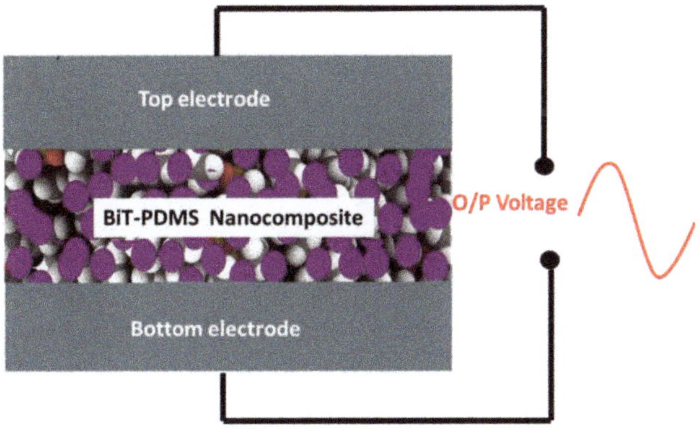

Figure 5.4. Schematic of a BIT-PDMS nanocomposite nanogenerator. Reproduced from [26]. Copyright (2017), with permission from Elsevier.

confirms the room-temperature multiferroicity in BIT. The addition of bismuth titanate in the PDMS polymer increased the internal polarization of the polymer, as demonstrated by the *P–E* loop. A simple device structure of the HNG device was constructed on the basis of the internal hybridization mechanism of the material BIT-PDMS composite used, which gave an electrical output of 300 V and 4.7 μA for a 16 cm^2 area device. The individual and multi-unit HNG devices prepared were compared for their performance at different frequencies and accelerations. The HNG device, along with the device structure, was mounted on a linear motor set-up to harvest the energy at different accelerations. The working mechanism of the HNG nanogenerator is shown in figure 5.5. When no external force is applied no electrical output or charge accumulation is observed (figure 5.5 (i)). When a force is applied, both the PDMS-BIT triboelectric layer and Al come in contact (figure 5.5 (ii)). Contact electrification is the reason behind the acquisition of positive and negative charge across the surfaces of the PDMS-BIT layer and Al electrode [28]. Moreover, the BIT particles present inside the PDMS-BIT composite film produce a piezoelectric potential due to resultant polarization [29]. This results in the development of a dipole moment due to the dislocation of charge centres of the cation and anion in bismuth titanate. To be precise, the piezoelectric potential developed across the BIT due to the applied force in addition to the triboelectric effect produces the combined effect and contributes to the total output of the HNG device. An instantaneous opposite piezoelectric potential is produced on the removal of the applied force. Concurrently, the negative charge existing on the surface of the PDMS-BIT film induces a positive electrostatic charge on the bottom Al electrode due to the electrostatic induction effect (figure 5.5 (iii)). Hence, the electrons move from the bottom Al electrode to the top Al electrode, producing a half-positive cycle. These electrons drift until the triboelectric charge dispersal reaches a steady state condition (i.e. returns to its initial state), as shown in figure 5.5 (iv). When the external force is again applied, the top triboelectric layer comes closer to the negative PDMS-BIT layer, facilitating the flow of electrons back to the bottom electrode, resulting in a negative cycle of the output signal (figure 5.5 (v)).

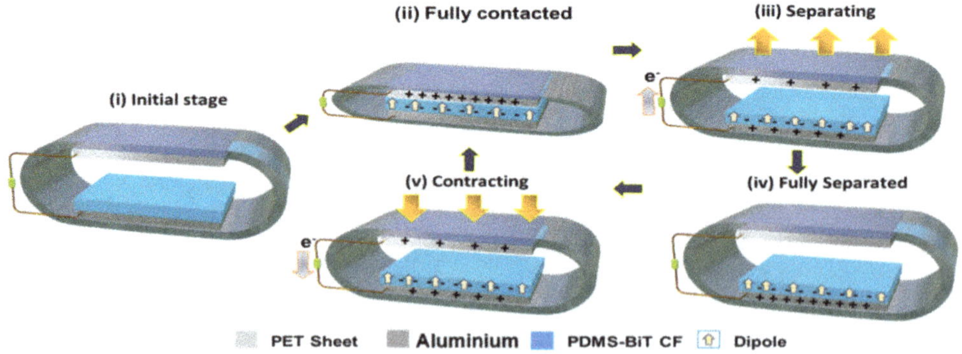

Figure 5.5. Working mechanism of HNG. Reproduced from [27]. Copyright (2021), with permission from Elsevier.

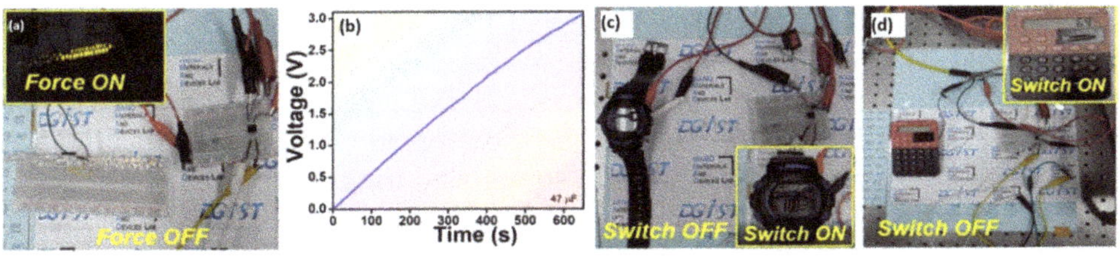

Figure 5.6. (a) The powering of LEDs using a bridge rectifier circuit. (b) Charging of the 47 μF capacitor by the 10 m s^{-2} acceleration for the powering of electronics. (c) An image of powering the wristwatch using the 47 μF capacitor. (d) An image of powering the calculator using the 47 μF capacitor. Reproduced from [27]. Copyright (2021), with permission from Elsevier.

Therefore, the alternating contact and separation of the prepared HNG device generates the electrical output owing to the synergy of the piezoelectric and triboelectric effects. Additionally, the HNG device can also be used to harvest biomechanical energy. The harvested energy can be effectively deposited inside a capacitor to substantially power low-power electronic devices such as LEDs, calculators, and wristwatches (shown in figure 5.6). The obtained results extend the application of hybrid piezo-triboelectric devices with high power density for several self-powered applications or blue energy harvesting.

Piezoelectric nanogenerators (PNGs) to be used as renewable energy sources were fabricated by Kim *et al* [30]. PNGs made up of organic piezoelectric materials such as polyvinylidene fluoride (PVDF) consisting of oxide complex powder have been a focus of study for various researchers for their flexible nature and high-performance energy conversion. Kim *et al* fabricated a PNG with a combination of PVDF and lanthanum-modified bismuth titanate ($Bi_{4-x}La_xTi_3O_{12}$, BLT) ceramics as typical ferroelectric materials. The BLT nano powders were obtained by treating the precursors with high-speed ball milling to reduce their particle size to the nanoscale. Kim *et al* also studied the particle size dependent energy harvesting performance of PNG without poling. PNG with BLT powders was fabricated by dispersing the BLT powder and PVDF in the N, N-dimethylacetamide (DMA) solution and stirred for 2 h at room temperature for homogenization. This solution was dropped on a soda-lime glass substrate placed in a spin coating unit rotating at an rpm of 1000 for 30 s. Then the coated samples were placed on a hot plate and dried at 60 °C for 3 h for solvent evaporation. The PVDF and BLT films were obtained by peeling the film from the glass substrate. PNG devices were then fabricated by sandwiching the films between two copper foils as the top and bottom electrodes. It was concluded that, due to the larger surface area of the nano-sized powder than that of the micro-sized powder, the nano-sized powder is uniformly distributed inside the PNG. Furthermore, the nano-sized powder-based PNG generates more than four times higher power than the PNG containing micro-sized powder.

5.7 Ultrasonic transducer

High temperature ultrasonic transducers made up of bismuth titanate were fabricated using a spray deposition technique combined with microwave sintering by Serfass *et al* [31]. Films with a thickness of 120 μm possessed a central frequency around 6 MHz. The elastic moduli, densities, dielectric constants, and conductivities of the as-prepared bismuth titanate films were evaluated. The piezoelectric coefficient d_{33} was found to be 16 pC N^{-1}. The as-prepared bismuth titanate films were found to have endurance for larger field strengths during poling, implying that the evaluated d_{33} value was accurate. The transducers fabricated by Serfass *et al* were proficient in producing ultrasound waves at 650 °C for a short period of time and showed that they were capable of operating at 600 °C for seven days and there was no evidence to suggest that they cannot operate at this temperature for still longer time periods. Additional experiments need to be undertaken to corroborate this.

5.8 Multifunctional applications

The various multifunctional properties of Bi-containing Aurivillius phase oxides of the $Bi_4Ti_3O_{12}$–$BiFeO_3$ binary system with a general formula of $Bi_{n+1}Fe_{n-3}Ti_3O_{3n+3}$ have attracted the interest of researchers, particularly in the last two decades. Benefiting from its high structural tolerance and co-existing electric dipole and magnetic ordering at room temperature, these Aurivillius phase oxides are potential single phase, room-temperature multiferroic materials [32].

Recent progress in the field of the synthesis of BFTO-n materials and the modification of their properties make them suitable for potential applications in the electronic industry, quantum devices, capacitors, transducers, microwave absorbers, catalysts, and photoelectric devices. Compared to $BiFeO_3$ and $BiFeO_3$-based materials, the BFTO-n materials not only possess giant polarization, a high Curie temperature, and RT magnetoelectric (ME) coexistence in the same phase, but also exhibit unique advantages, such as diversified modifications, larger ME effects (~400 mVOe^{-1} cm^{-1} at RT), low bandgaps (1.28–2.5 eV), RT single-phase exchange bias (EB) effects, microwave absorption, and excellent photo/electrocatalyst activity. In particular, considerable efforts have been made in the study of the ME multiferroic responses in BFTO-n. First, confirming the intrinsic multiferroic properties of BFTO-n is critical for boosting the performance and promoting its practical use in functional devices. Because of some controversies about the origin of the magnetism, intrinsic characterizations in ceramic bulks, thin films, and nanoparticles was addressed by both experimental and theoretical methods, in particular for the choice of preparation technique and the use of proper measurements. Second, there is the expectation of promoting FE and FM properties by substituting other cations into different sites as well as changing the number of perovskite units, because the manipulation of crystal structure can induce octahedral tilts and distortions, which largely influence the structural and physicochemical functionalities.

5.9 Photocatalysts

A lot of attention has been given to the Aurivillius oxides as potential candidates for highly active photocatalysts. The reasons for this are: (i) the presence of Bi enhances the likelihood of visible-light absorption by pushing the valence band edge due to the hybridization of the Bi 6s with the O 2p orbital; (ii) the largely dispersed hybridized valence band favours the high mobility of the photo-generated holes and is beneficial for the oxidation reaction; (iii) these materials possess narrow and tunable bandgaps; (iv) spontaneous polarization may be helpful in separating photon-generated electrons and hole pairs; and (v) ferromagnetism may be helpful in making recyclable catalysts. As they are better than the currently used selective TiO_2-based oxides mostly using UV light for photodegradation, the Aurivillius phase oxides may act as exceptional visible-light-driven photocatalysts, having wide applications not only in environmental remediation, water disinfection, pollutant degradation, air purification, and self-cleaning of surfaces, but also in the area of renewable energy generation in the form of hydrogen from water splitting [33–35].

5.10 Future scope of Aurivillius phase oxides

Based on the progress presented in the previous chapters, further research is still required to address the following challenges:

1. *Novel approaches to enhance properties*: Modifications of the micro/nano-structures and composition, such as various ion substitutions at different sites and variations of interfaces or layers, are to be established with an aim to improve the ferroelectric and ferromagnetic performance. However, each design approach results in limited enhancement of the performance. Therefore, developing new strategies is a significant and important subject with the help of the expansion of the family of Aurivillius phases containing various perovskite slabs. *In situ* control of the strain inside the layer structure may be a new strategy to realize the objective. Simultaneous symmetry and strain control, other perovskite-based derivatives, and phase-structure relations may be considered as methods to maximize the performance advantages to meet current demands. The study of quasi-continuous compositional changes across the homologous series may also be a possible approach.
2. *The development of ME coupling and ME devices*: Interestingly, some of the Aurivillius phase materials have already demonstrated ME coupling at or above room temperature along with the aptitude to be controlled by an externally applied electric or magnetic field. However, most of the reports concerning the ME effect are based on specific experimental interpretations. The comprehensive understanding of the relationship between the structure or layers and the ME effect of Aurivillius phase materials has been methodically investigated. The mechanism, mathematical modelling, theoretical simulation, and the effect of size on the ME response of Aurivillius phases still remains to be explored. The critical control of the Aurivillius thin films with exact composition and interfaces, and the proper magnetoelectric device design for applications are urgent requirements. Additionally, the Aurivillius based ME composites are also an important topic of study [36].
3. *New phenomena and functional applications*: Some attractive phenomena in the Aurivillius compounds, such as the exchange bias effect, microwave absorption, and photovoltaic effect, have been discovered. However, there is a long way to go for Aurivillius based materials to be used as microwave

absorbing materials and in photovoltaic devices, due to their low absorption capacity and low PV efficiency, respectively. In the meantime, these Aurivillius phase materials need to be further investigated for new phenomena or effects, such as local ordering [37], electro-caloric effects [38], Raman effects [39], etc, and functional properties in other areas, for example, oxide ion conductors [40], ultraviolet detectors [41], solid-oxide fuel cells [42], energy storage [43], drug carriers [44], and so on.

4. *Multiple catalytic activities*: The Aurivillius oxide compounds are regarded as a category of practical visible-light photocatalysts. However, it has been difficult to achieve high photocatalytic activity in Aurivillius based compounds for them to be suitable for catalytic applications. Construction of an Aurivillius oxide-based heterojunction might be the most effective way to further promote the photocatalytic activity. Moreover, due to the electrocatalytic activity, the electrochemical reduction of oxygen on the Aurivillius oxide catalysts was found. However, only a few reports focus on the mechanism for oxygen evolution reaction (OER), effects of catalyst characteristics, reaction rate, corrosion resistance, fabrication cost, and long-term stability of catalysts. Interestingly, due to the rich spectrum of the properties shown by this family of compounds, comprising ferroelectric polarization, piezoelectric activity, and pyroelectric effects, multiple catalytic activities can be taken into consideration in the future, specifically synergetic thermo-photo-catalysis [45] and sono-photo-catalysis [46], and could provide a major boost for the relevant technologies towards their desired commercialization.
5. *Other perovskite-based derivatives*: Perovskite-based derivation is a useful approach to induce remarkable properties or discover the fascinating phenomena of layered perovskites. Given the rich variety of structures, Aurivillius compounds have other perovskite-based derivatives in addition to S-A type bismuth-based layered oxyhalides. Layered structures exhibit the same basic perovskite motif but with different connecting layers. There are some inconsistencies in the connecting layers among these derivatives. A clear and general relationship between layer-stacked interfaces and properties needs to be established. This is essential for the design of Aurivillius based derivatives with tailored properties linked to different applications.

References

[1] Lencka M M, Oledzka M and Riman R E 2000 Hydrothermal synthesis of sodium and potassium bismuth titanates *Chem. Mater.* **12** 1323–30

[2] Lopatin S S, Lupeiko T G, Vasil'tsova T L, Basenko N I and Berlizev I M 1988 Properties of bismuth titanate ceramic modified with group V and VI elements *Inorg. Mater.* **24** 1328–31

[3] Chakraborty K R, Achary S N, Patwe S J, Krishna P S R, Shinde A B and Tyagi A K 2007 Low temperature neutron diffraction studies on $Bi_4Ti_3O_{12}$ *Ceram. Int.* **33** 601–4

[4] Ishiwara H, Okuyama M and Arimoto Y 2004 *Ferroelectric Random Access Memories: Fundamentals and Applications* vol 93 (Berlin: Springer)

[5] Prince B 2002 *Emerging Memories: Technologies and Trends* (Berlin: Springer)

[6] Moazzami R, Maniar P D, Jones R E, Campbell A C and Mogab C J 1993 Ferroelectric PZT thin films for low-voltage nonvolatile memory *Fifth Biennial Nonvolatile Memory Technology Review (June)* (Piscataway, NJ: IEEE) pp 44–7

[7] Noda M, Nomura S, Uchida H, Yamashita K and Funakubo H 2012 A new $SrBi_4Ti_4O_{15}$/$CaBi_4Ti_4O_{15}$ thin-film capacitor for excellent electric stability *IEEE Trans. Ultrason. Ferroelectr. Freq. Control* **59** 1888–93

[8] Guo F *et al* 2019 Evidence for the dynamic relaxation behavior of oxygen vacancies in Aurivillius Bi_2MoO_6 from dielectric spectroscopy during resistance switching *J. Mater. Chem.* C **7** 8915–22

[9] Kao M C, Chen H Z, Young S L, Chen K H, Chiang J L and Shi J B 2017 Structural, electrical, magnetic and resistive switching properties of the multiferroic/ferroelectric bilayer thin films *Materials* **10** 1327

[10] Megriche A, Lebrun L and Troccaz M 1999 Materials of $Bi_4Ti_3O_{12}$ type for high temperature acoustic piezo-sensors *Sens. Actuators* A **78** 88–91

[11] Song Y, Qu K, Zhao C, Ren J and Qu X 2010 Graphene oxide: intrinsic peroxidase catalytic activity and its application to glucose detection *Adv. Mater.* **22** 2206–10

[12] Yamauchi Y, Tonegawa A, Komatsu M, Wang H, Wang L, Nemoto Y, Suzuki N and Kuroda K 2012 Electrochemical synthesis of mesoporous Pt–Au binary alloys with tunable compositions for enhancement of electrochemical performance *J. Am. Chem. Soc.* **134** 5100–9

[13] Mao W, Cai B, Ye Z and Huang J 2018 A nanostructured p-NiO/n-$Bi_4Ti_3O_{12}$ heterojunction for direct GOx electrochemistry and high-sensitivity glucose sensing *Sens. Actuators* B **261** 385–91

[14] Zhang Z, Wang W, Wang L and Sun S 2012 Enhancement of visible-light photocatalysis by coupling with narrow-band-gap semiconductor: a case study on Bi_2S_3/Bi_2WO_6 *ACS Appl. Mater. Interfaces* **4** 593–7

[15] Pan C and Zhu Y 2010 New type of $BiPO_4$ oxy-acid salt photocatalyst with high photocatalytic activity on degradation of dye *Environ. Sci. Technol.* **44** 5570–4

[16] Bao M, Chen Y, Li F, Ma J, Lv T, Tang Y, Chen L, Xu Z and Wang T 2014 Plate-like p–n heterogeneous NiO/WO_3 nanocomposites for high performance room temperature NO_2 sensors *Nanoscale* **6** 4063–6

[17] Zhang Z, Shao C, Li X, Wang C, Zhang M and Liu Y 2010 Electrospun nanofibers of p-type NiO/n-type ZnO heterojunctions with enhanced photocatalytic activity *ACS Appl. Mater. Interfaces* **2** 2915–23

[18] Ratcliff E L, Meyer J, Steirer K X, Garcia A, Berry J J, Ginley D S, Olson D C, Kahn A and Armstrong N R 2011 Evidence for near-surface NiOOH species in solution-processed NiOx selective interlayer materials: impact on energetics and the performance of polymer bulk heterojunction photovoltaics *Chem. Mater.* **23** 4988–5000

[19] Wang C, Cheng X, Zhou X, Sun P, Hu X, Shimanoe K, Lu G and Yamazoe N 2014 Hierarchical α-Fe_2O_3/NiO composites with a hollow structure for a gas sensor *ACS Appl. Mater. Interfaces* **6** 12031–7

[20] Mao W, Cai B, Ye Z and Huang J 2018 A nanostructured p-NiO/n-$Bi_4Ti_3O_{12}$ heterojunction for direct GOx electrochemistry and high-sensitivity glucose sensing *Sens. Actuators* B **261** 385–91

[21] Wei C C, Yahaya M and Salleh M M 2005, July Nanocrystalline $Bi_4Ti_3O_{12}$ thin film for pressure sensor *Smart Sens. Actuators MEMS II* **5836** 380–8

[22] Yogo T, Kikuta K I, Yamada S and Hirano S I 1994 Synthesis of barium titanate/polymer composites from metal alkoxide *J. Sol-Gel Sci. Technol.* **2** 175–9
[23] Nakamura T, Muhammet R, Shimizu M and Shiosaki T 1995 Preparation of $Bi_4Ti_3O_{12}$ films by MOCVD and their application to memory devices *Integr. Ferroelectr.* **6** 35–46
[24] Fukuda K, Kitoh R and Awai I 1993 Microwave characteristics of TiO_2-Bi_2O_3 dielectric resonator *Jpn. J. Appl. Phys.* **32** 4584
[25] Zhang N, Zhang C and Li X J 2013 Preparation and properties of the ferroelectric materials based on BIT *Adv. Mater. Res.* **624** 146–9
[26] Abinnas N, Baskaran P, Harish S, Ganesh R S, Navaneethan M, Nisha K D, Ponnusamy S, Muthamizhchelvan C, Ikeda H and Hayakawa Y 2017 0.8 V nanogenerator for mechanical energy harvesting using bismuth titanate–PDMS nanocomposite *Appl. Surf. Sci.* **418** 362–8
[27] Hajra S, Padhan A M, Sahu M, Alagarsamy P, Lee K and Kim H J 2021 Lead-free flexible bismuth titanate–PDMS composites: a multifunctional colossal dielectric material for hybrid piezo-triboelectric nanogenerator to sustainably power portable electronics *Nano Energy* **89** 106316
[28] Kwon G D, Ahn B M, Lee J S, Park Y G and Ha Y C 2016 Clinical outcomes of a simultaneous replantation technique for amputations of four or five digits *Microsurgery* **36** 225–9
[29] Dudem B, Bharat L K, Patnam H, Mule A R and Yu J S 2018 Enhancing the output performance of hybrid nanogenerators based on Al-doped $BaTiO_3$ composite films: a self-powered utility system for portable electronics *J. Mater. Chem.* A **6** 16101–10
[30] Kim S, Nguyen T M H, He R and Bark C W 2021 Particle size effect of lanthanum-modified bismuth titanate ceramics on ferroelectric effect for energy harvesting *Nanoscale Res. Lett.* **16** 115
[31] Searfass C T, Pheil C, Sinding K, Tittmann B R, Baba A and Agrawal D K 2015 Bismuth titanate fabricated by spray-on deposition and microwave sintering for high-temperature ultrasonic transducers *IEEE Trans. Ultrason. Ferroelectr. Freq. Control* **63** 139–46
[32] Sun S and Yin X 2021 Progress and perspectives on Aurivillius-type layered ferroelectric oxides in binary $Bi_4Ti_3O_{12}$–$BiFeO_3$ system for multifunctional applications *Crystals* **11** 23
[33] Liu Y, Zhu G, Gao J, Hojamberdiev M, Zhu R, Wei X, Guo Q and Liu P 2017 Enhanced photocatalytic activity of $Bi_4Ti_3O_{12}$ nanosheets by Fe^{3+}-doping and the addition of Au nanoparticles: photodegradation of phenol and bisphenol A *Appl. Catalysis* B **200** 72–82
[34] Liu Z and Ma Z 2020 Promoting the photocatalytic activity of $Bi_4Ti_3O_{12}$ microspheres by incorporating iron *RSC Adv.* **10** 19232–9
[35] Bao L, Bao Q, Jiang W and Xu G 2020 Hydrothermal synthesis of Bi@ $Bi_4Ti_3O_{12}$ nanosheets with enhanced visible-light photocatalytic activity *Cryst. Eng. Comm.* **22** 6316–21
[36] Li D, Zhao X M, Zhao H X, Ren Y P, Long L S and Zheng L S 2020 Room-temperature magnetoelectric response in molecular–ionic ferroelectric-based magnetoelectric composites *Phys. Status Solidi—Rapid Res. Lett.* **14** 1900644
[37] Campanini M, Erni R and Rossell M D 2020 Probing local order in multiferroics by transmission electron microscopy *Phys. Sci. Rev.* **5** 20190068
[38] Chen J, Tang Z and Zhao S 2018 Giant negative and positive electrocaloric effects coexisting in lead-free $Na_{0.5}Bi_{4.5}Ti_4O_{15}$ films over a broad temperature range *Phys. Status Solidi—Rapid Res. Lett.* **12** 1700443
[39] Rodríguez Aranda M D C, Rodríguez-Vázquez Á G, Salazar-Kuri U, Mendoza M E and Navarro-Contreras H R 2018 Raman effect in multiferroic $Bi_5Fe_{1+x}Ti_{3-x}O_{15}$ solid solutions: a temperature study *J. Appl. Phys.* **123** 084101

[40] Kendall K R, Navas C, Thomas J K and zur Loye H C 1996 Recent developments in oxide ion conductors: Aurivillius phases *Chem. Mater.* **8** 642–9

[41] Pintilie L, Alexe M, Pignolet A and Hesse D 1998 $Bi_4Ti_3O_{12}$ ferroelectric thin film ultraviolet detectors *Appl. Phys. Lett.* **73** 342–4

[42] Zhu Y, Zhou W, Chen Y and Shao Z 2016 An Aurivillius oxide based cathode with excellent CO_2 tolerance for intermediate-temperature solid oxide fuel cells *Angew. Chem. Int. Ed.* **55** 8988–93

[43] Tang Z, Chen J, Yang B and Zhao S 2019 Energy storage performances regulated by layer selection engineering for doping in multi-layered perovskite relaxor ferroelectric films *Appl. Phys. Lett.* **114** 163901

[44] Nair M, Guduru R, Liang P, Hong J, Sagar V and Khizroev S 2013 Externally controlled on-demand release of anti-HIV drug using magneto-electric nanoparticles as carriers *Nat. Commun.* **4** 1–8

[45] Xu X, Xiao L, Jia Y, Wu Z, Wang F, Wang Y, Haugen N O and Huang H 2018 Pyro-catalytic hydrogen evolution by $Ba_{0.7}Sr_{0.3}TiO_3$ nanoparticles: harvesting cold–hot alternation energy near room-temperature *Energy Environ. Sci.* **11** 2198–207

[46] Li X, Yin X, Gu W, Zhu L, Liu H, Zhang G, Fu Z and Lu Y 2018 Sonocatalysis of the magnetic recyclable layered perovskite oxides *Ultrason. Sonochem.* **49** 260–7

www.ingramcontent.com/pod-product-compliance
Ingram Content Group UK Ltd.
Pitfield, Milton Keynes, MK11 3LW, UK
UKHW051851100926
8556IPUK00005B/565

9 780750 334174